CHAOS
IN SYSTEMS WITH NOISE

CHAOS
IN SYSTEMS WITH NOISE

Tomasz Kapitaniak
Technical University of Lodz,
Lodz, Poland

World Scientific
Singapore • New Jersey • Hong Kong

Published by

World Scientific Publishing Co. Pte. Ltd.
P.O. Box 128, Farrer Road, Singapore 9128

U. S. A. office: World Scientific Publishing Co., Inc.
687 Hartwell Street, Teaneck NJ 07666, USA

Library of Congress Cataloging-in-Publication Data

Kapitaniak, Tomasz.
Chaos in systems with noise.

Bibliography: p.
1. Chaotic behavior in systems. 2. Random noise theory.
3. Stochastic processes. I. Title.
Q172.5.C45K37 1988 003 88-1294
ISBN 9971-50-542-8

Printed by Sun U Book Co. Sdn. Bhd., Petaling Jaya.

PREFACE

In this book the influence of the random noise on the chaotic behaviour of dissipative dynamics systems is investigated. We generally do not deal with discrete - time dynamics systems e.g., iterative maps, but we consider continuous - time dynamics. The problems presented here are illustrated by some mechanical examples.

The main part of this book is based on author's post-doctoral thesis "Chaotic stochastic processes in non-linear dynamics" which has been presented at the Technical University of Lodz, but it also presents the results which are the combined achievements of many investigators.

The author is very grateful to Prof. W.-H. Steeb of the Rand Afrikaans University in Johannesburg and Prof. A. Tylikowski of Warsaw Technical University for their valuable suggestions during the discussion of many problems treated in this book.

CONTENTS

CHAOS
IN SYSTEMS WITH NOISE

1. INTRODUCTION - CHAOTIC AND STOCHASTIC PROCESSES AND THEIR KOLMOGOROV ENTROPY

First let's consider the dynamics system of the general type:

$$\bar{x} = \bar{f}(\bar{x}, t) \tag{1.1}$$

where $\bar{x} = [x_1, \ldots, x_n]^T$ represent the variables characterizing the temporal evolution of the system in n-dimensional phase space, $\bar{f} = [f_1, \ldots, f_n]^T$ gives the coupling between different variables $\bar{f}$ is assumed to be differentiable with respect to $\bar{x}$.

The temporal evolution of the system (1.1) can be represented by the evolution of a probability distribution $p(t)$ in n-dimensional phase space. We consider an initial distribution $p_0 = p(t_0)$. As it is impossible to represent a state of the system as a point in phase space, either due to fundamental uncertainty relations or due to the principal inability of dealing with infinitely precise values of phase space coordinates, a finite volume

$$\delta V = \delta x_1 \ldots\ldots \delta x_n$$

in phase space has to be associated with p_0 .

We are interested in the temporal evolution of the system with respect to some small uncertainties in the initial condition p_0 , so we consider the temporal evolution of the uncertainties δx_i forming δV :

$$\frac{d}{dt} \delta x_i(t) = \sum_j \frac{\partial \bar{f}_i(\bar{x}(t))}{\partial x_j} \delta x_j(t) \tag{1.2}$$

the coefficients $\frac{\partial f(\bar{x}(t))}{\partial x_j}$ form a matrix L , leading to

Lyapunov exponents λ_i which will be described in Chapter 7 .

The approximate solution of equation (1.2) is of the form:

$$\delta x_i(t) = \delta x_i(t_o) \exp(\lambda_i (t - t_o)) \tag{1.3}$$

The temporal evolution of the volume element defined by p_o is independent of a choice of the coordinate system and is given by :

$$\delta V(t) = \delta V(t_o) \exp(\sum_{i=1}^{n} \lambda_i (t - t_o)) \tag{1.4}$$

(Atmanspacher and Scheingraber (1987)) .

The above equation gives a simple criterion to distinquish between conservative and dissipative systems. For $\delta V(t) = 0$, the volume of a solution in phase space is conserved and this is the case of conservative systems. (We have also

$$\sum_{i=1}^{n} \lambda_i = 0$$

for conservative systems.) In dissipative systems the phase space volume $\delta V(t) < 0$, which is realized by

$$\sum_{i=1}^{n} \lambda_i < 0$$

Lyapunov exponents are independent of the initial condition (Oseledec (1968)) so if the system is dissipative or conservative

$$\sum_{i=1}^{n} \lambda_i < 0 ,$$

the evolution of the system takes place in a limited subspace of the phase space. The specific subspace which is asymptotically reached in time is called the <u>attractor</u> of the system. In case of

$$\sum_{i=1}^{n} \lambda_i > 0$$

the system may never reach any attractor.

Lyapunov exponent helps in the characterization of different types of attractors of dynamics system.

If we have one-dimensional phase space, the only possible attractor $\lambda_1 = 0$ is represented by a point in phase space.

In a two-dimensional phase space, one has to consider two Lyapunov exponents. The combination $(\lambda_1, \lambda_2) = (-, -)$ provides a point like attractor and the combination $(\lambda_1, \lambda_2) = (0, -)$ is visualized by a limit cycle in the phase space which corresponds to the periodic solution.

For a three-dimensional system there are the following stable solutions:

(- , - , -) point - like attractor

(0 , - , -) limit cycle , periodicity

(0 , 0 , -) torus , periodicity two frequencies

In addition, there are non-trivial attractors with (+ , 0 , -) whenever

$$\sum_{i=1}^{n} \lambda_i < 0$$

(according to Poincare - Bendixson theorem - Arnold (1980)) these types of attractors cannot occur in less than three dimensional phase space).

Attractors with positive Lyapunov exponents are called strange attractors . As the equation (1.1) is completely deterministic, strange attractors are purely deterministic.

The sum of positive Lyapunov exponents describes the rate of information production of the system. Under a certain restriction (Procaccia (1985) , Ruelle (1979)), this rate can be identified with the Kolmogorov entropy H :

$$H = \sum_{i} \begin{cases} \lambda_i & \text{if } \lambda_i > 0 \\ 0 & \text{otherwise} \end{cases} \tag{1.5}$$

The above definition states, that for solutions with $\lambda_i \leqslant 0$ for all i the Kolmogorov entropy H = 0 . This condition characterized the class of regular processes (periodic, almost periodic). Chaotic solutions require $H > 0$, because at least one Lyapunov exponent has to be greater than 0 .

Now let's consider pure stochastic system:

$$\dot{x}(t) = \bar{\eta}(\bar{x}, t) \tag{1.6}$$

where $\bar{\eta} = [\bar{\eta}_1, \ldots, \bar{\eta}_n]^T$ is a stochastic process with zero mean and correlation function:

$$\langle \eta_i(t) \ \eta_j(t') \rangle = D_i \, \delta(t - t') \, \delta_{ij} \tag{1.7}$$

where δ is Dirac's delta , δ_{ij} is Kronecker delta , D_i

are constants.

In this case $\bar{x}(t)$ is random variable obeying a probability distribution in phase space. At a time t_o a state of the system (1.6) can be determined with inevitable uncertainties forming $\delta V(t_o)$. For an infinitely small advanced time $t=t_o+dt$, for at least one $x_i(t)$ there exists a nonvanishing probability which is situated in an interval $\delta x_i(t) > \delta x_i(t_o)$. From this, we may conclude that the uncertainty $\delta x_i(t_o)$ is expanding immediately after t_o, and due to the description in formula (1.4) λ_i has to be infinite in this situation and the same is with Kolmogorov entropy ($H = \infty$ for pure stochastic process).

The above results of Kolmogorov entropy are summarized in Table 1.1.

	Kolmogorov entropy
regular	$H = 0$
chaotic	$H > 0$
stochastic	$H = \infty$

Since nature does not provide processes without fluctuations random perturbation, the system (1.1) can be treated only as a pure idealization.

Stochastic perturbation can be taken into account by considering the following system:

$$\dot{\bar{x}}(t) = \bar{f}(\bar{x},t) + \bar{\eta}(\bar{x}, t) \qquad (1.8)$$

The superposition of regular and stochastic process has been investigated by many authors, for example: Tylikowski and Skalmierski (1982), Piszczek and Nizioł (1986), Roberts and Spanos (1986), Tagata (1978), Stratanovich (1963).

The problem of the superposition of chaotic and stochastic processes, especially in dissipative systems, is relatively a new one (Crutchfield and Huberman (1980), Crutchfield et al. (1982)).

In this book some methods of investigating these problems are presented. These methods allow to distinquish between chaotic and regular processes perturbed by noise. We do not deal with discrete-time dynamics systems. The problem of noise influence on chaos in such dynamics system has been investigated by: Argoul et al. (1987), Herzel et al. (1987), Herzel and Pompe (1987), Guckenheimer (1982), Kapral et al. (1982), Thomas and Grossmann (1981) Zipplius and Lucke (1981), Ott and Hanson (1981), Sharaiman et al. (1981), Crutchfield et al. (1981), Valle et al. (1984), Ott et al. (1985).

2. NOISY DYNAMICS SYSTEMS

Consider noisy dynamics system described by the following system of n stochastic differential equations:

$$\frac{\partial(\bar{x},t,\omega)}{\partial t} = \bar{f}(\bar{x}, t, \omega) \tag{2.1}$$

with initial condition:

$$\bar{x}(0,\omega) = [x_1(0,\omega), \ldots, x_n(0,\omega)]^T = [x_{10}, \ldots, x_{n0}]^T$$

where $\bar{x} = [x_1,\ldots,x_n]^T$, $\bar{f} = [f_1,\ldots,f_n]^T$, $\omega \in \Omega$, $(\Omega, \mathcal{B}, \mu)$ is a probabilistic space where Ω, $\mathcal{B}$ and μ are respectively the set of elementary events, the σ-field of its Borel subsets, the probabilistic measure.

In the next chapters we shall be particularly interested in the realizations of the solution (response of the system 2.1) so we must assume that the solution response of noisy dynamics system $\bar{x}(t,\omega)$ is differentiable with probability 1.

Definition 2.1

Consider probabilistic space $(\Omega, \mathcal{B}, \mu)$, time interval $T = [t_o, b)$ $b \leqslant \infty$, function $F\colon T \times R^n \times \Omega \rightarrow R^n$ and initial value $\bar{x}(0,\omega) = x_o(\omega)$.

If there exists constant δ and the function $x : [t_o, \delta) \times \Omega \rightarrow R^n$ which is differentiable in the interval $[t_o, \delta)$ for nearly all $\omega \in \Omega$ them $\bar{x}$ fulfils equation (2.1), initial condition $\bar{x}(0,\omega) = \bar{x}_o$ and $\bar{x}$ is called the almost sure solution of initial value problem almost sure response of the noisy dynamics system.

Besides the above mentioned definition of the almost sure solution also another definitions of the solution of the stochastic initial value problem were introduced, for example:

Definition 2.2

If for all conditions of definition 2.1 we have for any $t \in [t_0, \delta)$

$$P(\dot{\bar{x}}(t,\omega) = \bar{F}(t, \bar{x}, \omega)) = 1$$

the stochastic process $\bar{x}(t,\omega)$ is called the locally almost sure solution of initial value problem.

or

Definition 2.3

Random process $\{\bar{x}(t,\omega),\ t \in [t_0,\delta),\ \omega \in \Omega\}$ is called a stochastic solution of the stochastic initial value problem if:

1. $\bar{x}(t,\omega)$ belongs to the class of all n-dimensional random vectors of $(\Omega, \mathcal{B}, \mu)$ for all $t \in [t_0, \delta) = T$
2. For all $t \in T$, each $\varepsilon > 0$ and any $\eta > 0$ there exists $\varrho(t, \varepsilon, \eta)$, such that from $|h| < \varrho(t, \varepsilon, \eta)$ we have:

$$P\Big\{\omega \in \Omega : |\, h^{-1} [\ \bar{x}(t+h,\omega) - \bar{x}(t,\omega)] - \bar{F}(t, \bar{x}(t,\omega))\ | > \eta\Big\} < \varepsilon$$

If one solution is the almost sure solution, it is also locally almost sure and stochastic solution, but the oposite implication fails.

From the above mentioned definition 2.1 we learn that the process $\{x(t,\omega)\ ,\ t\in[t_o,\delta) = T\}$, is the almost sure solution of the initial value problem almost sure response of the noisy dynamics system , when nearly all its realizations in the interval T are the solutions of deterministic initial value problem (for settled $\omega\in\Omega$).

For the existence and uniqueness of the almost sure solution we have the following theorem:

Theorem 2.1

Let $F: T \times R^n \times \Omega \rightarrow R^n$ fulfil the following conditions:

1. $F(t, x,\omega)$ belongs to the class of all n-dimensional vectors of $(\Omega, \mathcal{B}, \mu)$
2. For all $\omega\in\Omega$ function $F_\omega(\cdot, \cdot, \cdot) : T\times R^n \rightarrow R^n$ is continuous with the exception of the subset $H_\omega \subset T$ of measure 0 , which does not contain any finite points of accumulation
3. For nearly all $\omega\in\Omega$ and all $x\in R^n$ the function $F_\omega(\cdot,x)$: $T \rightarrow R^n$ is bounded in all finite intervals $T_x \subset T$
4. There exists the function $L: T\times\Omega \rightarrow R_+$ continous on T , for nearly all $\omega\in\Omega$, that function is such that for nearly all $\omega\in\Omega$, all $t\in T$ and $x, y \in R^n$

$$|F_\omega(t, x) - F_\omega(t, y)| \leqslant L(t,\omega)\ |x - y|$$

For all the above conditions the unique almost sure solution of the stochastic initial value problem 2.1 excists.

Proof : Existence

Let ω be one-settled elementary event, which fulfils the assumptions of the theorem.

Let $x_k : T \times \Omega \to R^n$ $(k=0,1,2,\ldots)$ be a sequence of functions, which approximate the solution of the stochastic initial value problem:

$$x_0(t,\omega) = x_0(\omega)$$
$$x_k(t,\omega) = x_0(\omega) + \int_{t_0}^{t} F_\omega[s, x_{k-1}(s,\omega)]\,ds \quad (k=1,2..) \tag{2.2}$$

As $x_0(t,\omega)$ is a continuous function of time , from the assumption 2 we get that the function $F_\omega[\,\cdot\,, x_0(\cdot,\omega)]$ is continuous on $T - X_\omega$.

Also from the fact that $x_0(t,\omega)$ is continuous we have that all functions $x_k(t,\omega)$ are continuous for nearly all $\omega \in \Omega$.

Let ω be still settled and let $[a, b] \subset T$ be an interval which contains t_0 $(t_0 \in [a, b])$. From assumption 4 we have for each $t \in [a, b]$ and $k \geqslant 0$:

$$|x_{k+1}(t,\omega) - x_k(t,\omega)| = \left| \int_{t_0}^{t} \{F_\omega[s, x_k(s,\omega)]\right.$$

$$\left. - F_\omega[s, x_{k-1}(s,\omega)]\}ds \right| \leqslant \int_{t_0}^{t} | F_\omega[s, x_k(s,\omega)]$$

$$- F_\omega[s, x_{k-1}(s,\omega)]|\,ds \leqslant \int_{t_0}^{t} L(s,\omega)\ |x_k(s,\omega)$$

$$- x_{k-1}(s,\omega)|\,ds \leqslant \sup_{t\in[a,b]} L(t,\omega) \int_{t_0}^{t} |x_k(s,\omega) - x_{k-1}(s,\omega)|\,ds$$

and for each $k=0,1,\ldots$

$$|x_{k+1}(t,\omega) - x_k(t,\omega)|$$

$$\leqslant \int_{t_0}^{t} \dots \sup_{s\in[a,b]} L(s,\omega) \int_{t_0}^{t} \sup_{s\in[a,b]} L(s,\omega) \int_{t_0}^{t} |F[s, x_0(\omega)]| ds dt_1 \dots dt_k$$

$$\leqslant \frac{(b-a)^{k+1}}{(k+1)!} \left[\sup_{s\in[a,b]} L(s,\omega) \right]^k \sup_{s\in[a,b]} | F_\omega [s, x_0(\omega)] |$$

From theory of deterministic differential equations for example (Arnold (1983)) and from the fact that series:

$$\sum_{k=0}^{\infty} \frac{(b-a)^{k+1}}{(k+1)!} \left[\sup_{s\in[a,b]} L(s,\omega) \right]^k \sup_{s\in[a,b]} | F_\omega [s, x_0(\omega)]$$

is convergent we conclude that the sequence of functions $x_k(t,\omega)$ $(k=0,1,\dots)$ converges to $\overline{x}(t,\omega) : T \rightarrow R^n$ which fulfils deterministic differential equation:

$$\dot{\overline{x}}(t,\omega) = F_\omega [t, \overline{x}(t,\omega)] \qquad t\in T - H_\omega$$

and initial condition $\overline{x}(t_0,\omega) = \overline{x}_0(\omega)$.

By the appropriate selection of an interval $[a, b]$, it is possible to construct solution $\overline{x}(t,\omega)$ for all $t\in T$ and all $\omega \in \Omega_1 \subset \Omega$ $(P(\Omega_1) = 1)$. Hence, we have functions $\{\overline{x}(t,\omega), (t,\omega) \in T \times \Omega_1\}$ which fulfil initial value problem (2.1) for settled ω. Now we have to prove that these functions are the realizations of the stochastic process $\{\overline{x}(t,\omega), t\in T, \omega\in\Omega\}$

As x_0 is the random variable from the theory of measurable random functions (Doob (1953)) we get that there exists $\mathcal{B}\otimes\mathcal{T}_n$ measurable function $G(t,\cdot)$ for which:

$$G_\omega (t, x) = F_\omega (t, x) \qquad (t, x) \in T \times R^n$$

So, the integrals in the formula (2.2) are as the limits of Riemann's sum $\mathcal{B}\otimes\mathcal{T}_n$ measurable and by induction it is possible to show the existence of the stochastic process $\{\overline{x}_{t,k}, t\in T\}$ for which $\overline{x}_{t,k}(\omega) = x_k(t,\omega)$ for $t \in T$.
As

$$\lim_{k\to\infty} x_k(t,\omega) = x(t,\omega)$$

we can conclude that there exists $\{\overline{x}_t, t\in T\}$ which fulfils the condition $\overline{x}_t(\omega) = x(t,\omega)$ and which is an almost sure solution of the initial value problem (2.1).

Uniqueness

If two stochastic processes $\overline{x}(t,\omega)$, $t\in T, \omega\in\Omega$ and $\overline{y}(t,\omega)$, $t\in T$, $\omega\in\Omega$ are the solutions of the stochastic initial value problem (2.1) from the assumption 4 we obtain, that for all $t\in T$ and nearly all

$$|\overline{x}(t,\omega) - \overline{y}(t,\omega)| \leqslant \int_{t_0}^{t} L(s,\omega)\ |\overline{x}_s(\omega) - \overline{y}_s(\omega)|\,ds$$

and $P[|\overline{x}(t,\omega) - \overline{y}(t,\omega)| = 0] = 1$, so stochastic processes $\overline{x}(t,\omega)$ and $\overline{y}(t,\omega)$ are stochastically equivalent.

3. FOKKER - PLANCK - KOLMOGOROV EQUATION

Let's start with the definition of Markov's process:

Definition 3.1

Let $x_1(t_1), \ldots, x_n(t_n)$ be the sequence of values of the random process $x(t,\omega)$ at $t_1 < t_2 < \ldots < t_n$.

The process $x(t,\omega)$ is a Markovian process if the conditional probability density at t_n depends only on the last value $x_{n-1}(t_{n-1})$, so the following relationship holds:

$$P_c(x_n, t_n \mid x_{n-1}, t_{n-1}, \ldots, x_1, t_1)$$

$$= P_c(x_n, t_n \mid x_{n-1}, t_{n-1}) \tag{3.1}$$

$$= \frac{P_n(x_1, \ldots, x_n, t_1, \ldots, t_n)}{P_{n-1}(x_1, \ldots, x_{n-1}, t_1, \ldots, t_{n-1})} .$$

In particular, the conditional probability density fulfils the normalizing condition :

$$\int_{-\infty}^{\infty} P_c(x, t \mid x_1, t_1)\, dx = 1 \tag{3.2}$$

and Smoluchowski's equation :

$$\int_{-\infty}^{\infty} P_c(x, t \mid x_1, t_1)\, P_c(x_1, t_1 \mid x_0, t_0)\, dx_1$$

$$= P_c(x, t \mid x_0, t_0) \tag{3.3}$$

The Smoluchowski's equation for t_0 can be considered

as representing n-dimensional Markovian process $\overline{x}(t,\omega) = [x_1(t,\omega), x_2(t,\omega), \ldots, x_n(t,\omega)]^T$. In this case we have:

$$P(\overline{x}, t+\Delta t \mid x_o, 0) = \int_{-\infty}^{\infty}\!\!\cdots\!\int P(\overline{x}, t+\Delta t \mid \overline{y}, t)\; P(\overline{y}, t \mid \overline{x}_o, 0)\; dy_1 dy_2 \ldots dy_n \tag{3.4}$$

For the sake of simplicity the index c has been omitted in (3.4) and in the following ones.

Multiplying both sides of the above equation by any scalar function $Q(\overline{x})$, which is developed into Taylor's series and which converges sufficiently rapidly to zero with its derivatives when $x \to \pm\infty$ and integrating both sides of the obtained expression, we get

$$\int_{-\infty}^{\infty}\!\!\cdots\!\int Q(\overline{x}) P(\overline{x}, t+\Delta t \mid \overline{x}_o, 0)\, dx_1 \ldots dx_n$$

$$= \int_{-\infty}^{\infty}\!\!\cdots\!\int dy_1 \ldots dy_n \int_{-\infty}^{\infty}\!\!\cdots\!\int Q(\overline{x}) P(\overline{x}, t+\Delta t \mid y, t) \tag{3.5}$$

$$P(\overline{y}, t \mid \overline{x}_o, 0)\, dx_1 \ldots dx_n$$

Expanding the function $Q(\overline{x})$ with Taylor's series in the vicinity of $\overline{y}$ we obtain:

$$Q(\overline{x}) = Q(\overline{y}) + \sum_{i=1}^{n} (x_i - y_i) \left.\frac{\partial Q}{\partial x_i}\right|_{x_i = y_i}$$

$$+ \frac{1}{2} \sum_{i=1}^{n} \sum_{j=1}^{n} (x_i - y_i)(x_j - y_j) \left.\frac{\partial^2 Q}{\partial x_i\, \partial x_j}\right|_{x_k = y_k} \tag{3.6}$$

By substituting (3.6) into the right side of (3.5) we obtain:

$$\int_{-\infty}^{\infty}\!\!\cdots\!\int dy_1\ldots dy_n \;\Big[\, Q(\overline{y})P(\overline{y},\; t|\overline{x}_o,\, 0)$$

$$\int_{-\infty}^{\infty}\!\!\cdots\!\int P(\overline{x},\; t+\Delta t|\,\overline{y},\; t)\, dx_1\ldots dx_n \tag{3.7}$$

$$+\sum_{i=1}^{n} \frac{\partial Q}{\partial x_i}\bigg|_{x_i=y_i} P(\overline{y},t|\overline{x}_0,0)\int_{-\infty}^{\infty}\!\!\cdots\!\int (x_i-y_i)\, P(\overline{x},t+\Delta t|\,\overline{y},t)\, dx_1\ldots dx_n$$

$$+\sum_{i=1}^{n}\sum_{j=1}^{n} \frac{\partial^2 Q}{\partial x_i \partial x_j}\bigg|_{x_k = y_k} P(\overline{y},\; t|\overline{x}_0,0)$$

$$\int_{-\infty}^{\infty}\!\!\cdots\!\int (x_i - y_i)(x_j - y_j)\;\; P(\overline{x},\; t+\Delta t|\,\overline{y},\; t)\;\, dx_1\ldots dx_n + \ldots$$

$$= \int_{-\infty}^{\infty}\!\!\cdots\!\int \Big[\, Q(\overline{x})P(\overline{x},t|\overline{x}_0,0) + \sum_{i=1}^{n} A_i(\overline{x},t,\Delta t)\,\frac{\partial Q}{\partial x_i}\; P(\overline{x},\; t|\overline{x}_0,0)$$

$$+\frac{1}{2}\sum_{i=1}^{n}\sum_{j=1}^{n} B_{ij}(\overline{x},t,\Delta t)\,\frac{\partial^2 Q}{\partial x_i\partial x_j}\; P(\overline{x},\; t|\overline{x}_0,0) + \ldots \Big]\, dx_1\ldots dx_n$$

where

$$A_i(\overline{x},\; t,\; \Delta t) = \langle (x_i - y_j)\rangle$$

$$= \int_{-\infty}^{\infty}\!\!\cdots\!\int (x_i - y_i)\, P(\overline{x},\ t+\Delta t|\,\overline{y},\ t)\, dx_1 \ldots dx_n \qquad (3.8)$$

$$B_{ij}(\overline{x},\ t,\ \Delta t) = \langle (x_i - y_i)\ (x_j - y_j) \rangle$$

$$= \int_{-\infty}^{\infty}\!\!\cdots\!\int (x_i - y_i)(x_j - y_j)\, P(\overline{x},\ t+\Delta t|\,\overline{y},\ t)\, dx_1 \ldots dx_n \qquad (3.9)$$

denotes mean value and where the expression

$$\int_{-\infty}^{\infty}\!\!\cdots\!\int P(\overline{x},\ t+\Delta t|\,\overline{y},\ t)\, dx_1 \ldots dx_n = 1 \qquad (3.10)$$

is used.

Integrating equation (3.7) piecewise and applying the result (3.5) we get:

$$\int_{-\infty}^{\infty}\!\!\cdots\!\int Q(\overline{x}) \Big\{ P(\overline{x},\ t+\Delta t|\,\overline{x}_o,\ 0) - P(\overline{x},\ t|\,\overline{x}_o,\ 0)$$

$$+ \sum_{i=1}^{n} \frac{\partial}{\partial x_i} [\, A_i P(\overline{x},\ t|\,\overline{x}_o,\ 0)] \qquad (3.11)$$

$$- \frac{1}{2} \sum_{i=1}^{n} \sum_{j=1}^{n} \frac{\partial^2 Q}{\partial x_i \partial x_j}\ [\, B_{ij} P(\overline{x},\ t|\,\overline{x}_o, 0)] + \ldots \Big\} dx_1 \ldots dx_n = 0$$

If both sides of this equation are divided by t and if they approach the limit where $\Delta t \rightarrow 0$, we obtain:

$$\frac{\partial P}{\partial t} + \sum_{i=1}^{n} \frac{\partial}{\partial x_i} [a_i P] - \frac{1}{2} \sum_{i=1}^{n} \sum_{j=1}^{n} \frac{\partial^2}{\partial x_i \partial x_j} [b_{ij} P] = 0 \tag{3.12}$$

The above equation (3.12) is called the Fokker - Planck - Kolmogorov equation. The initial condition for this equation is as follows:

$$P(\overline{x}, 0 | \overline{x}_0, 0) = \delta(\overline{x} - \overline{x}_0) = \prod_{i=1}^{n} \delta(x_i - x_{io}) \tag{3.13}$$

The transition from (3.11) to (3.12) is justified by the fact that $Q(\overline{x})$ is an arbitrary function of the arguments $x_1, \ldots, x_n$ and $Qa_i P \to 0$ when $\overline{x} \to \pm\infty$. The coefficients a_i and b_{ij} in equation (3.12) are as follows:

$$a_i(\overline{x}, t) \tag{3.14}$$

$$= \lim_{\Delta t \to 0} \frac{1}{\Delta t} \int_{-\infty}^{\infty} \cdots \int (y_i - x_i) P(\overline{y}, t+\Delta t | x, t)\, dy_1 \ldots dy_n$$

and

$$b_{ij}(\overline{x}, t) \tag{3.15}$$

$$= \lim_{\Delta t \to 0} \frac{1}{\Delta t} \int_{-\infty}^{\infty} \cdots \int (y_i - x_i)(y_j - x_j) P(\overline{y}, t+\Delta t | \overline{x}, t)\, dy_1 \ldots dy_n$$

assuming that limits exist. It is also assumed that the remaining expressions connected with Taylor's series in (3.7) become zero when $\Delta t \to 0$.

Finally let's go back to the stochastic differential equation (2.1). In this book we shall be particularly interested in the following form of this equation:

$$\frac{dx_j(t,\omega)}{dt} = f_j(x_1,\dots,x_n) + \eta_j(t,\omega) \quad (j=1,2,\dots,n) \tag{3.16}$$

Now we show that the almost sure solution of the above equation is a Markovian process when $\eta_j(t,\omega)$ are stationary, statistically independent white noise of zero mean value and correlation function:

$$\langle \eta_i(t,\omega)\, \eta_j(t',\omega)\rangle = D_j\, \delta_{ik}\, \delta(t-t') \tag{3.17}$$

$$i,\ j = 1,2,\dots,n$$

where D_j is a variance of the process and δ_{ik} and δ are respectively Kronecker and Dirac's deltas.

Equation (3.16) denotes the Markovian process $[x_1(t,\omega), x_2(t,\omega), \dots, x_n(t,\omega)]$ for its probability density, providing that $x_i(t,\omega)$ $(i=1,2,\dots,n)$ for $t \leqslant t_o$ depends only on the value $x_i(t_o) = x_{io}$ since

$$x_j(t,\omega) = x_{jo} + \int_{t_o}^{t} f_j(x_1,\dots,x_n)\, d\tau + \int_{t_o}^{t} \eta(\tau,\omega)\, d\tau \tag{3.18}$$

does not depend on time up to moment t_o.

Based on equation (3.16) the increments of the following function can be calculated:

$$\Delta x_j = f_j(\overline{x})\, \Delta t + \int_{t}^{t+\Delta t} \eta_j(\tau,\omega)\, d\tau \tag{3.19}$$

where for the sake of simplicity we put $\overline{x} = [x_1, \ldots, x_n]^T$.

Substituting expression (3.19) into (3.8)

$$A_j = f_j(\overline{x})\Delta t + \int_t^{t+\Delta t} \eta_j \langle(\tau,\omega)\rangle \, d\tau \qquad (3.20)$$

is obtained and on the basis of (3.14) and (3.19) we get:

$$a_j(x_1, \ldots, x_n) = f_j(x_1, \ldots, x_n) \qquad (3.21)$$

The coefficients b_{ij} are calculated in the following way. From (3.19) we get:

$$\Delta x_i \, \Delta x_j = [f_i \Delta t + \int_t^{t+\Delta t} \eta_i \, (\tau_1,\omega) \, d\tau_1][f_j \Delta t \qquad (3.22)$$

$$+ \int_t^{t+\Delta t} \eta_j \, (\tau_2,\omega) \, d\tau_2] = f_i f_j \, (\Delta t)^2$$

$$+ f_j \Delta t + \int_t^{t+\Delta t} \eta_i \, (\tau_1,\omega) \, d\tau_1 + f_i \Delta t \int_t^{t+\Delta t} \eta_j \, (\tau_2,\omega) \, d\tau_2$$

$$+ \int_t^{t+\Delta t} \int_t^{t+\Delta t} \eta_i \, (\tau_1,\omega) \, \eta_j \, (\tau_2,\omega) \, d\tau_1 d\tau_2$$

Substituting (3.22) into (3.9) we obtain:

$$B_{ij} = \langle \Delta x_i \, \Delta x_j \rangle = f_i f_j \, (\Delta t)^2 \qquad (3.23)$$

$$+ \int_t^{t+\Delta t} \int_t^{t+\Delta t} \langle \eta_i \, (\tau_1,\omega) \, \eta_j \, (\tau_2,\omega) \rangle \, d\tau_1 d\tau_2$$

$$= f_i f_j (\Delta t)^2 + D_j \Delta t$$

From equation (3.14)

$$b_{ij}(x_1, \dots, x_n) = D_i \ \delta_{ij} \qquad (3.24)$$

Analytical solution of F-P-K equations is difficult to determine. In some cases it is easier to find the solution of F-P-K equations in the stationary state when

$$\frac{\partial P}{\partial t} = 0 \ .$$

In this book all probability density functions have been calculated numerically.

4. MULTI-MAXIMA PROBABILITY DENSITY FUNCTIONS

4.1 Stochastic process with bifurcation

Investigations of the probability density function of the response of the noisy non-linear systems show that this function can be characterized not only by one maxima curve but also by multi-maxima curves.

The examples of amplitude probability density functions and waveforms connected with them are shown in Figure 4.1

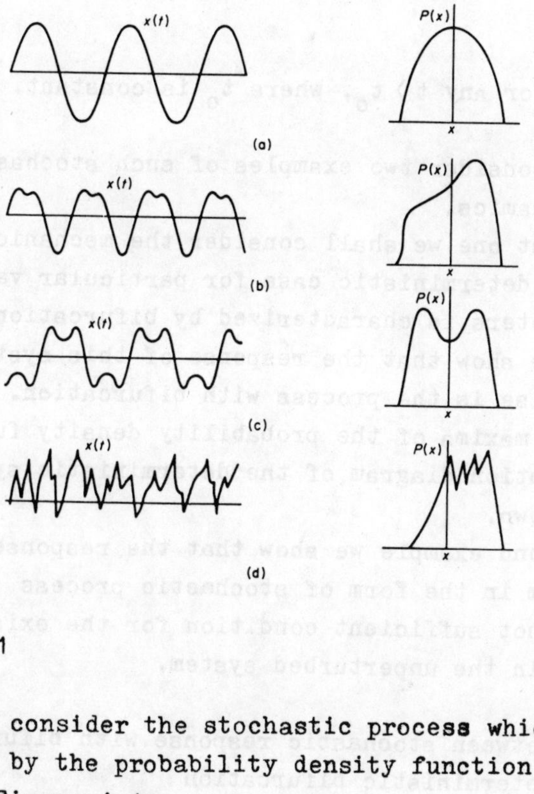

Figure 4.1

First consider the stochastic process which is characterized by the probability density function with two maxima - Figure 4.1c

Definition 4.1

Let $\overline{x}(t,\omega) = [x_1(t,\omega),\ldots,x_n(t,\omega)]^T$ be a stochastic process given by equation (3.15) with a probability density function $P(x_1,\ldots,x_n,t|x_{10},\ldots,x_{n0})$.

Stochastic process $x_i(t,\omega)$ $i=1,2,\ldots,n$ is called the process with bifurcation if the probability density function $P(x_i,t|x_{i0})$ given by equation:

$$P(x_i, t|x_{i0}) \tag{4.1}$$

$$= \int_{-\infty}^{\infty}\!\!\cdots\!\int P(x_1,\ldots,x_n,t|x_{10},\ldots,x_{n0})\,dx_1\ldots dx_{i-1}dx_{i+1}\ldots dx_n$$

has two maxima for any $t > t_o$, where t_o is constant.

Now let's consider two examples of such stochastic processes in dynamics.

In the first one we shall consider the mechanical system which in deterministic case for particular values of system parameters is characterized by bifurcation of the solution. We show that the response of this system perturbed by noise is the process with bifurcation. The analogy between maxima of the probability density function and bifurcation diagram of the deterministic system will be also shown.

In the second example we show that the response of the noisy system in the form of stochastic process with bifurcation is not sufficient condition for the existence of bifurcation in the unperturbed system.

4.2 Relation between stochastic response with bifurcation and deterministic bifurcation

4.2.1 Non-linear pendulum perturbed by white noise

Consider mechanical system shown in Figure 4.2 (Thompson 1982).

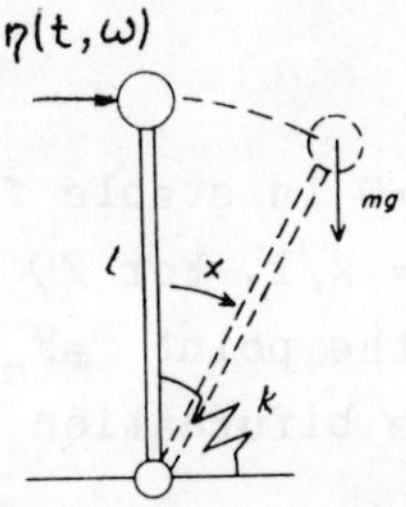

Figure 4.2 : Model of the system. k, torsional stiffness; m, mass; l, length of the massless beam.

This system is perturbed by stationary random force with zero mean and correlation function:

$$\langle \eta(t,\omega) \quad \eta(t',\omega)\rangle = D\,\delta(t - t')$$

The potential energy of the system is as follows:

$$V = \frac{1}{2} kx^2 - Fl(1 - \cos x) \qquad (4.2)$$

By equating derivative of it to zero:

$$\frac{\partial V}{\partial x} = k - Fl\sin x = 0 \qquad (4.3)$$

we obtain a trivial solution $x = 0$ for any F and the solutions given by the equation:

$$F = \frac{kx}{l\sin x} \qquad (4.4)$$

From the investigations of the second derivative of the potential energy:

$$\frac{\partial^2 V}{\partial x^2} = k - Fl\cos x \tag{4.5}$$

we learn that the solution x=0 is stable for $F < F_c$ and unstable for $F > F_c$ where $F_c = k/l$. For $F > F_c$ the solution given by (4.4) is stable. In the point $F=F_c$ we have a bifurcation of the solution. The bifurcation diagram x versus F is shown in Figure 4.3

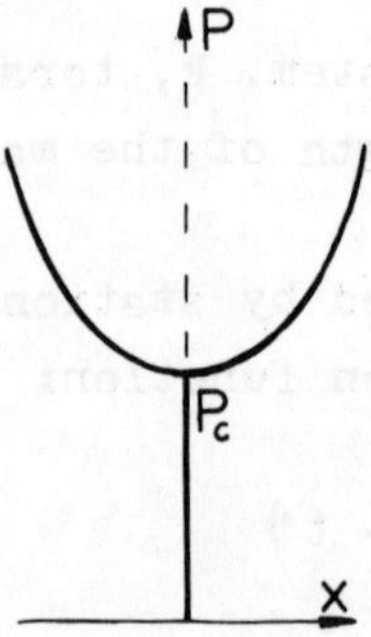

Figure 4.3

Now let's consider the influence of an additive random noise $\eta(t,\omega)$ on the behavicur of the system.

The equation of motion of the body with mass m can be described in the following way:

$$\begin{aligned} \dot{x}_1 &= x_2 \\ \dot{x}_2 &= ax_1 - \lambda \sin x_1 + \eta(t,\omega) \end{aligned} \tag{4.6}$$

where : $x_1 = x$, $x_2 = \dot{x}$, $a = k/ml^2$ and $\lambda = F/ml$.

Equation (4.6) is a particular example of equation (3.15)

The stationary state $(\partial P/\partial t = 0)$ F-P-K equation for this system is as follows:

$$- \frac{\partial}{\partial x_1} x_2 P - \frac{\partial}{\partial x_2}[(ax_1 - \lambda \sin x_1)P] + \frac{D}{2}\frac{\partial^2 P}{\partial x_2^2} = 0 \qquad (4.7)$$

where $P(x_1, x_2 | x_{10}, x_{20}, \lambda)$,(x_{10} and x_{20} are initial conditions) is a probability density function in the stationary state .

The calculated probability density functions of the displacement x_1 for different values of the bifurcation parameter λ ,

$$P(x_1, \lambda) = \int_{-\infty}^{\infty} P(x_1, x_2, \lambda)\, dx_2 \qquad (4.8)$$

are presented in Figure 4.4

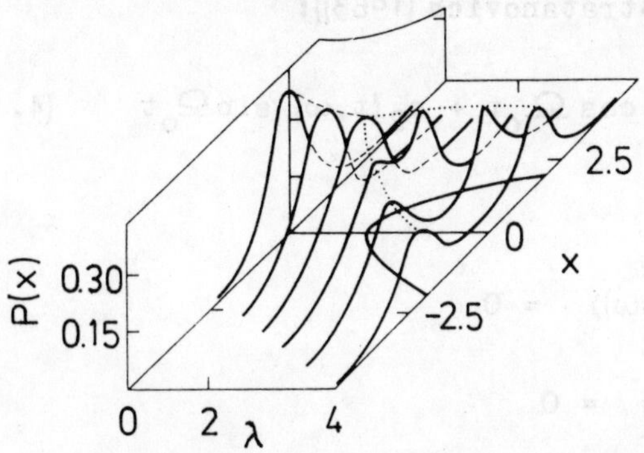

Figure 4.4 : The analogy between stochastic process with bifurcation and the bifurcation of the deterministic system ($\eta(t,\omega) = 0$): a=2.0, D=0.5

The projection of the maxima of this function on the amplitude-bifurcation parameter plane gives bifurcation diagram of deterministic system known from Figure 4.3

4.2.2. Duffing's equation perturbed by large narrow-band noise

As the second example consider oscillations of an anharmonic oscillator:

$$\ddot{x} + a\dot{x} + bx + cx^3 = \varepsilon \eta\,(t,\omega) \qquad (4.9)$$

where $\bar{\eta}(t,\omega)$ is large narrow-band noise obtained by passing white noise $\eta(t,\omega)$ thought linear filter with central frequency Ω_o:

$$\ddot{\bar{\eta}}(t,\omega) + \alpha\dot{\bar{\eta}}(t,\omega) + \Omega_o^2\,\bar{\eta}(t,\omega) = \eta\,(t,\omega) \qquad (4.10)$$

where α is the damping factor (Tagata (1978)).

Stochastic process $\bar{\eta}\,(t,\omega)$ can be described in the following form (Stratanovich (1963)):

$$\bar{\eta}(t,\omega) = n_1\,(t,\omega)\ \cos\Omega_o t + n_2\,(t,\omega)\ \sin\Omega_o t \qquad (4.11)$$

where:

$$\langle n_1\,(t,\omega)\rangle = \langle n_2\,(t,\omega)\rangle = 0$$

$$\langle n_1\,(t,\omega)\ \ n_2\,(t';\omega)\rangle = 0$$

$$\langle n_1\,(t,\omega)\ n_1\,(t';\omega)\rangle = \langle n_2\,(t,\omega)\ n_2\,(t';\omega)\rangle = D\ \ \delta(t - t')$$

Equation (4.9) has been solved for different realizations of random process $\eta(t,\omega)$ by Runge-Kutta method for stochastic equations. Probability density functions have been estimated.

Depending on the values of system parameters we obtain different shapes of probability density function - Figure 4.5 . For the following values of parameters :

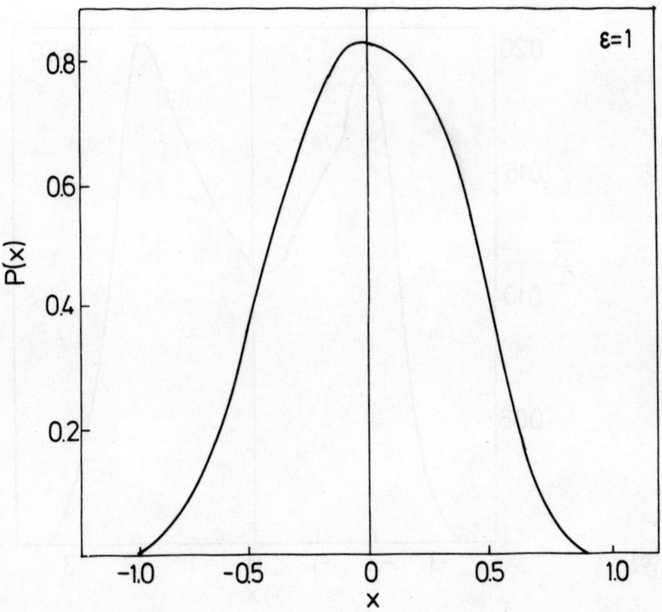

(a)

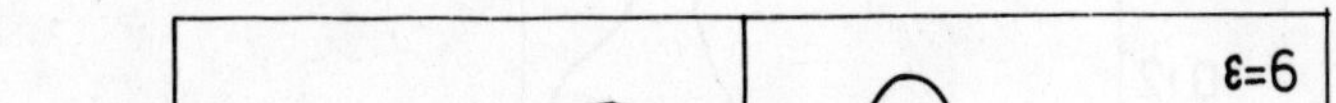

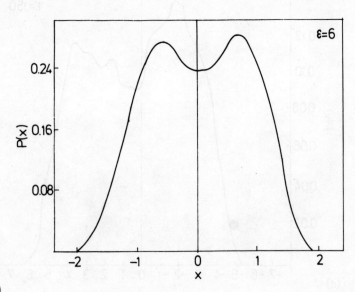

(b)

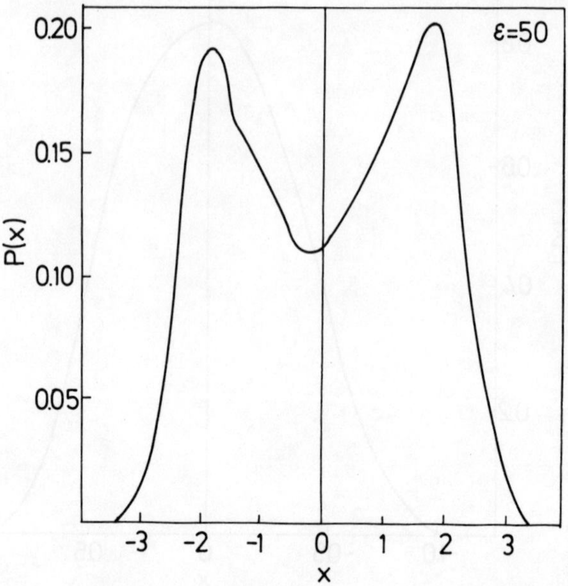
0.20
0.15
0.10
0.05
P(x)
ε=50
-3 -2 -1 0 1 2 3
x

(c)

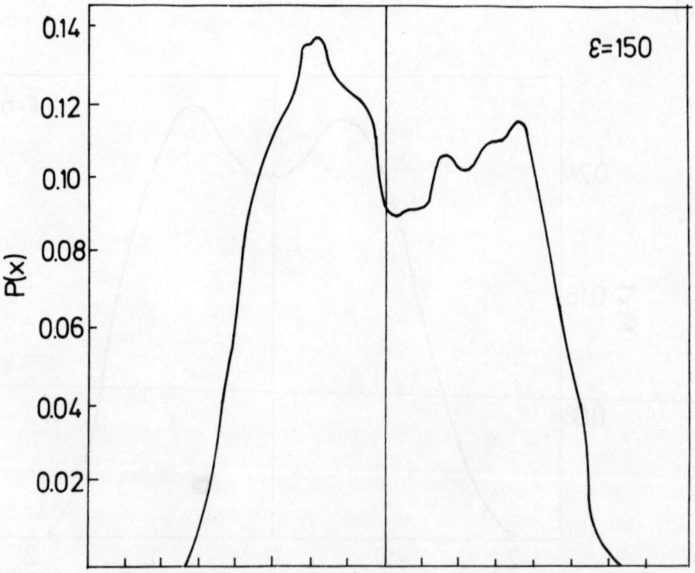
0.14
0.12
0.10
0.08
0.06
0.04
0.02
P(x)
ε=150
-7 -6 -5 -4 -3 -2 -1 0 1 2 3 4 5 6 7
x

(d)

Figure 4.5

α =0.2 , a=0.4 , b=0.9 , c=3.6 and $\varepsilon \in [6, 76]$ we obtain amplitude probability density function with two maxima. The examples of realizations of the process $x(t,\omega)$ are shown in Figure 4.6

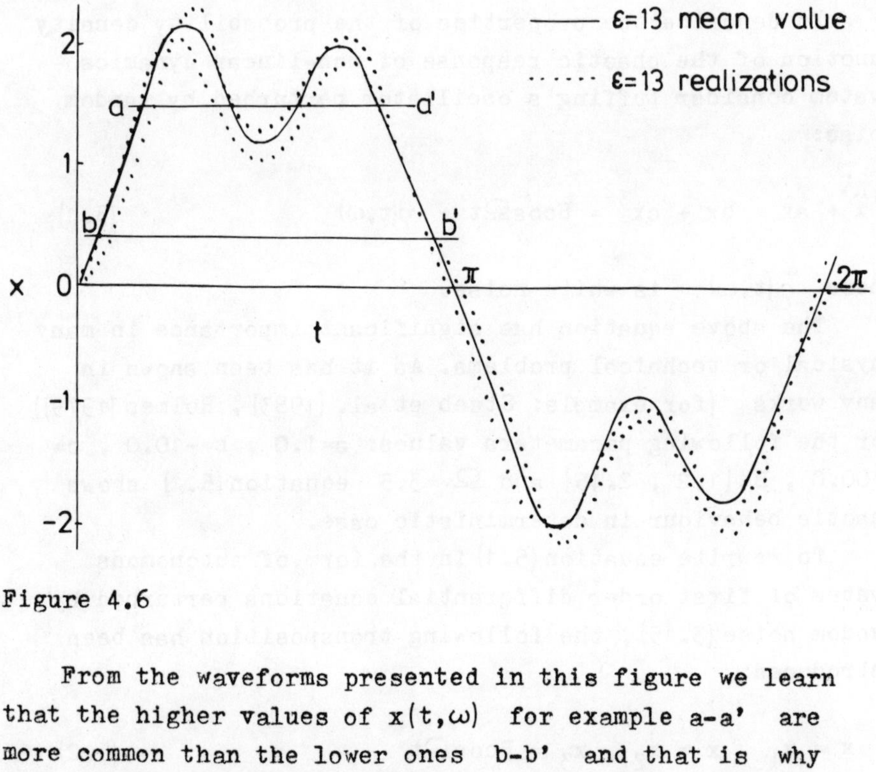

Figure 4.6

From the waveforms presented in this figure we learn that the higher values of $x(t,\omega)$ for example a-a' are more common than the lower ones b-b' and that is why the probability density functions - P(x) have two maxima.

For the same as the above values of parameters α , a, b and c , $\varepsilon > 76$ the probability density function has more than two maxima. In this case the process $x(t,\omega)$ can be called a stochastic process with multifurcation.

This example shows that the stochastic response with bifurcation can take place also in the systems which in deterministic case are not characterized by bifurcation of the solution.

5. CHAOTIC AND REGULAR STOCHASTIC PROCESSES

5.1. Properties of the probability density function of the noisy chaotic response

To describe the properties of the probability density function of the chaotic response of non-linear dynamics system consider Duffing's oscillator perturbed by random noise:

$$\ddot{x} + a\dot{x} + bx + cx^3 = B\cos\Omega t + \eta(t,\omega) \qquad (5.1)$$

where $\eta(t,\omega)$ is white noise.

The above equation has significant importance in many physical or technical problems. As it has been shown in many works (for example: Steeb et al. (1983), Holmes (1979)) for the following parameters values: $a=1.0$, $b=-10.0$, $c=100.0$, $B\in[1.2, 2.45]$ and $\Omega=3.5$ equation (5.1) shows chaotic behaviour in deterministic case.

To rewrite equation (5.1) in the form of autonomous system of first order differential equations perturbed by random noise (3.15), the following transposition has been introduced:

$$x = x_1 , \; x = x_2 , \; x_3 = B\cos\Omega t \qquad (5.2)$$

where x_3 is the solution of the following initial value problem:

$$\begin{aligned} \dot{x}_3 &= x_4 \\ \dot{x}_4 &= -\Omega^2 x_3 \end{aligned} \qquad (5.3)$$

with initial values: $x_3(0) = B$, $x_4(0) = 0$.

After this transposition we get:

$$\dot{x}_1 = x_2$$

$$\dot{x}_2 = - ax_2 - bx_1 - cx_1^3 + x_3 + \eta (t,\omega)$$

$$\dot{x}_3 = x_4 \tag{5.4}$$

$$\dot{x}_4 = - \Omega^2 x_3$$

With the probability density function denoted by $P(x_1, x_2, x_3, x_4, t | x_{10}, x_{20}, B, 0)$, where x_{10} and x_{20} are the initial conditions of the system (5.1) the F-P-K equation satysfying equations (5.4) becomes:

$$\frac{\partial P}{\partial t} = - \frac{\partial}{\partial x_1} [x_2 P] - \frac{\partial}{\partial x_2} [(- ax_2 - bx_1 - cx_1^3 + x_3) P] - \frac{\partial}{\partial x_3} [x_4 P] - \frac{\partial}{\partial x_4} [-\Omega^2 x_3 P] + \frac{D}{2} \frac{\partial^2 P}{\partial x_2^2} \tag{5.5}$$

The probability density of $x(t,\omega)$ can be calculated from the integral:

$$P(x_1, t) = \iiint_{-\infty}^{\infty} P(x_1, x_2, x_3, x_4, t | x_{10}, x_{20}, B, 0)\, dx_2 dx_3 dx_4 \tag{5.6}$$

In Figure 5.1 the examples of the probability density function of the response $x(t,\omega)$ for settled constant time t are shown.

For the system parameters for which the system show chaotic behaviour in deterministic case we obtain probability density functions with multi maxima - Figure 5.1 (c) and (d).

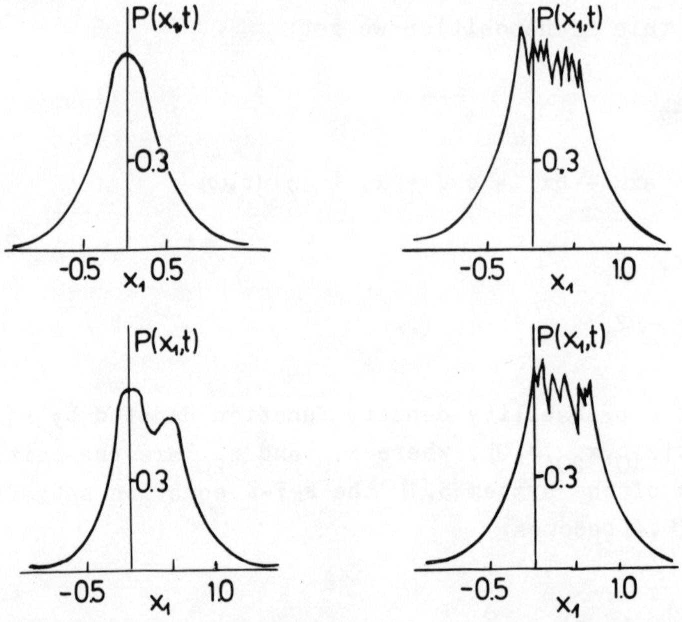

Figure 5.1 : Types of the probability density function of the oscillator (5.1) : a=1.0 , b=-10.0 , c=100.0 B=1.2 , D=0.2 : (a) Ω =3.2 , (b) Ω =3.4 , (c) and (d) Ω =3.5 , (a) - (c) t=20.0 , (d) t=40.0

The comparison of the value of probability density function for constant value of x and different values of time - Figure 5.1 (c) and (d), shows that this value is not constant in time. The examples of plot P(x, t) for constant x versus time have been shown in Figure 5.2 .

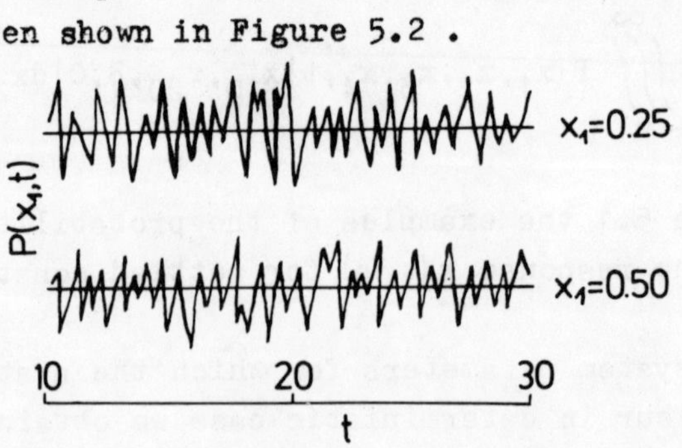

Figure 5.2

5.2 Definition of chaotic stochastic process

As it was shown in the previous point the probability density function with multi-maxima is characteristic for chaotic behaviour of the noisy system.

The multi-maxima curves occur due to the fact that in the oscillation waveform there exist values of the amplitude which are more probable than neighbouring values. For example in Figure 5.3 the amplitudes a-a' and c-c' occur more often than amplitude b-b' .

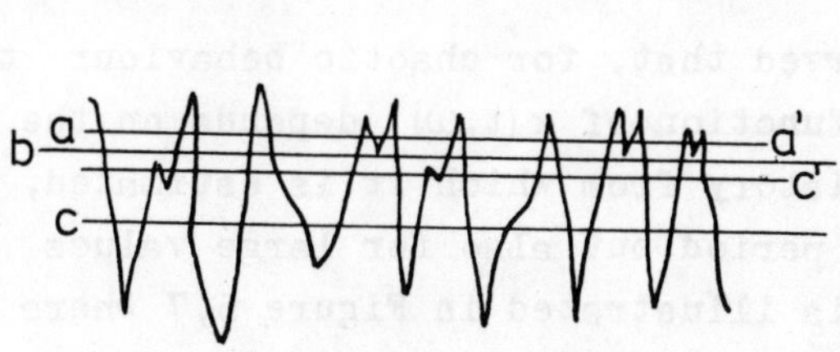

Figure 5.3

On the other hand a multi-maxima curve is not a sufficient indicator of chaotic behaviour. A similar multi--maxima probability density function has been obtained for the system (5.1) at a=0.1 , b=0 , c=1.0 , B=10.0 and Ω =1.03 , as is shown in Figure 5.4

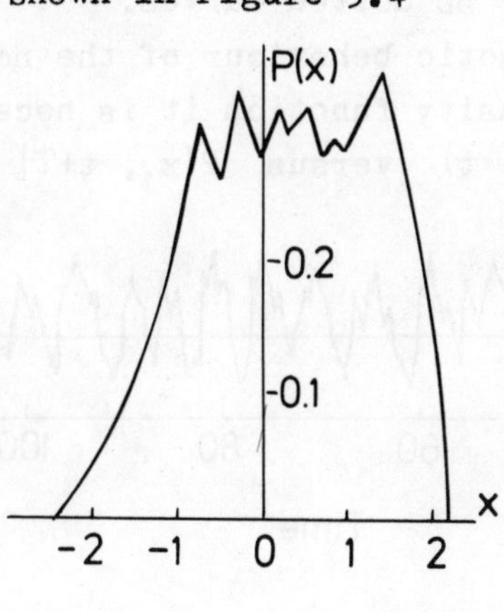

Figure 5.4

In this case the system does not possess a chaotic behaviour but an almost periodic one, as it has been shown on the Poincare map in Figure 5.5

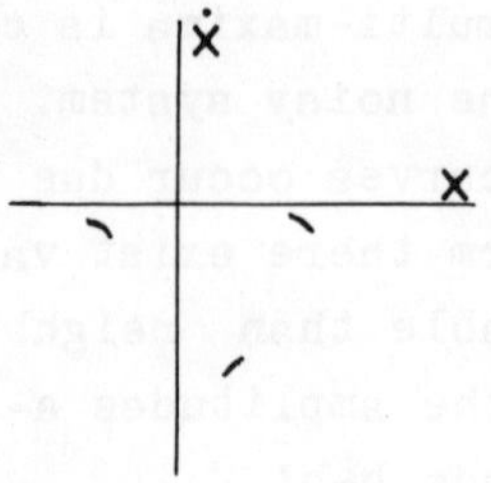

Figure 5.5

It has been observed that, for chaotic behaviour the probability density function of $x(t,\omega)$ depends on the length of the time history from which it is estimated, not only for the initial period but also for large values of time. This property is illustrated in Figure 5.7 where based on the time history of Figure 5.6 of the chaotic behaviour of the system(5.1) at a=0.1 , b=0 , c=1.0 , B=10.0 and Ω=1.0 (see Ueda 1979 , Szemplińska-Stupnicka 1986), the probability density functions have been estimated with time period lengths of 80s and 120s , and different shapes of curves have been obtained. The same property can be obtained from F-P-K equation (5.5). These results have been shown in Figure 5.7 as dotted curves.

To identify chaotic behaviour of the noisy system from the probability density function it is necessary also to make a map of $P(x_1, t)$ versus $P(x_1, t+\tau)$ for constant

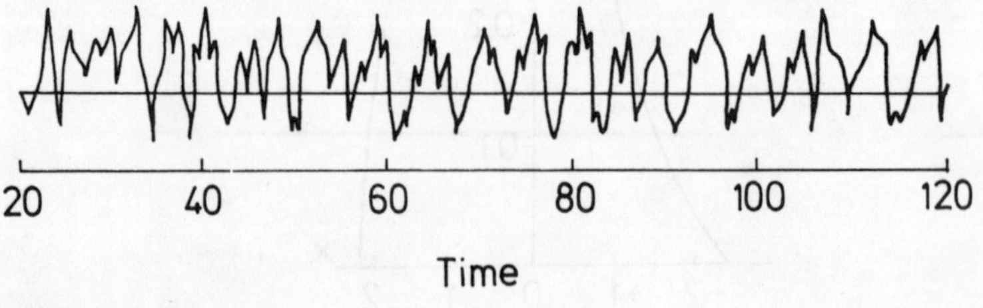

Figure 5.6

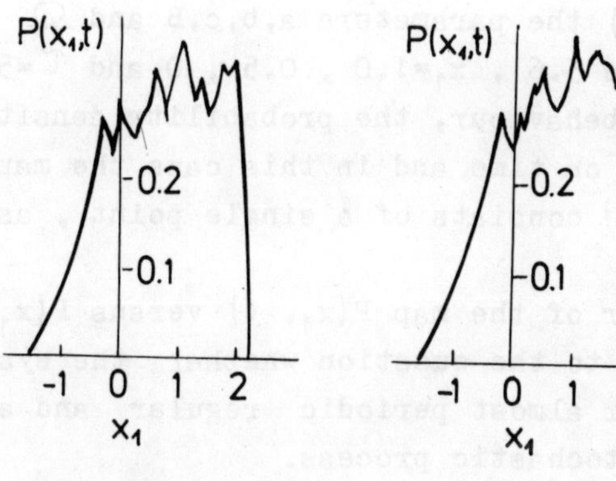

Figure 5.7 : (a) t=80.0 , (b) t=120.0

x_1 and τ . The examples of such maps are shown in Figure 5.8 . In the case of chaotic behaviour, the numerical results indicate that the maps have a Cantor set structure,

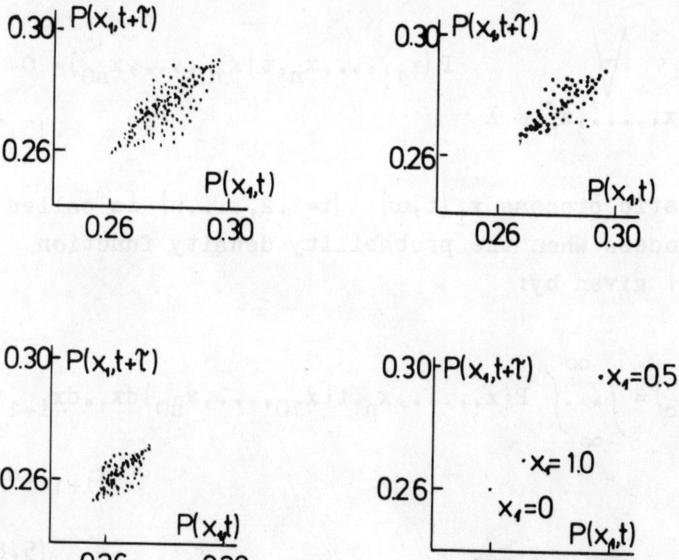

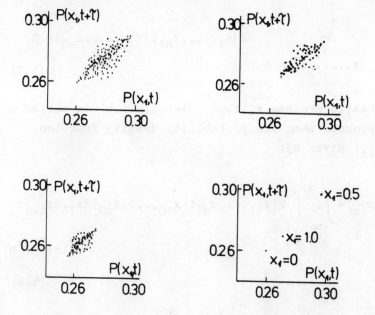

Figure 5.8 : (a) x_1=1.0 , (b) x_1=0.5 , (c) x_1=0 , (a) - (c) chaotic behaviour, (d) regular behaviour, τ=5.

Figure 5.8 (a) - (c) the parameters a,b,c,B and Ω are the same as in Figure 5.6 , $x_1=1.0$, 0.5 , 0 and $\tau=5$.

For regular behaviour, the probability density function does not depend on time and in this case the map $P(x_1, t)$ versus $P(x_1, t+\tau)$ consists of a single point , as shown in Figure 5.8 (d).

The character of the map $P(x_1, t)$ versus $P(x_1, t+\tau)$ provides the answer to the question whether the system behaviour is chaotic or almost periodic regular and allows to define chaotic stochastic process.

Definition 5.1

Let $\overline{x}(t,\omega) = [x_1(t,\omega),\ldots,x_n(t,\omega)]^T$ be a stochastic process given by equation (3.15) with probability density function $P(x_1,\ldots,x_n,t|x_{10},\ldots,x_{n0})$.

Let

$$\underset{X\subset R^n}{\exists} \quad \underset{(x_1,\ldots,x_n)\notin X}{\forall} \quad P(x_1,\ldots,x_n,t|x_{10},\ldots,x_{n0}) = 0 \tag{5.7}$$

Stochastic process $x_i(t,\omega)$ $(i=1,2,\ldots,n)$ is called a chaotic process when the probability density function $P(x_i,t|x_{i0})$ given by:

$$P(x_i,t|x_{i0}) = \int_{-\infty}^{\infty}\!\!\cdots\!\int P(x_1,\ldots,x_n,t|x_{10},\ldots,x_{n0})\,dx_1\ldots dx_{i-1}\times \\ \times\, dx_{i+1}\ldots dx_n \tag{5.8}$$

is a function with multi-maxima and

$$\exists_{x_i=x_o} \quad \forall_{\tau>0} \quad \overline{\overline{\{(P(x_o,\ t), P(x_o,\ t+\tau)) \mid t \in [0,\infty)\}}} = \aleph_o \qquad (5.9)$$

where $\overline{\overline{\{\cdot\}}}$ indicates power of the set and $\aleph_o$ is power of the natural numbers set.

Remark

The condition (5.7) indicates that the response of the system (3.15) is biunded:

$$\exists_{M_1,\ldots,M_n} \quad \forall_{t} \quad x_1(t,\omega) < M_1,\ldots,\ x_n(t,\omega) < M_n \qquad (5.10)$$

The defined above chaotic stochastic process is a nonstationary process, but the class of chaotic processes is narrower than the class of nonstationary processes. For example the response of the system:

$$\ddot{x} - a\dot{x} + bx = \eta\ (t,\omega)$$

where $a, b > 0$ and $\eta(t,\omega)$ is stationary process is nonstationary, but it is not chaotic (the condition (5.7) is not fulfilled).

Definition 5.2

The same as in definition 5.1 the stochastic process $x_i(t,\omega)$ is called a regular one when the probability density function $P(x_i, t|x_{io})$ fulfils the following condition:

$$\forall_{x_i = x_o} \quad \exists_{\tau>0} \quad \overline{\overline{\{(P(x_o,t)\ ,\ P(x_o,\ t+\tau)) \mid t \in [0,\infty)\ \}}} < \aleph_o$$

6. POINCARE MAPS FOR NOISY SYSTEMS

6.1. Definition of Poincare map

The theoretical base for Poincare maps has been introduced by Poincare (see Poincare (1898), Mardsen and McCracken (1976)). The development of the computational methods and determination of the chaotic behaviour in dynamics systems (Lorentz (1963), Ueda (1979)) cause that the method of Poincare map becomes one of the popular and the most illustrative method of describing "strange attractor".

During investigations of the dynamics system we are particularly interested in the asymptotic behaviour of the phase trajectories. This allows us to investigate the behaviour of the phase trajectory points of the specially selected time periods. The Poincare map consists of these points.

Now, let's present the mathematical definition of the Poincare map for ordinary differential equations (Mardsen and McCracken (1976)).

First of all we recall that a closed orbit γ of a flow F_t (F_t is the flow of a C^k vector field X) on a manifold M is a non-constant integral curve $\gamma(t)$ of X such that $\gamma(t+\tau) = \gamma(t)$ for all $t \in R$ and some $\tau > 0$. (The least such τ is the period of γ). The image of γ is clearly diffeomorphic to a circle.

Definition 6.1

Let γ be a closed orbit, let $m \in \gamma$, see $m = \gamma(0)$ and let S be a local transversal section, (submanifold of codimension one transverce to γ i.e. $\gamma'(0)$ is not tangent to S). Let $\mathcal{D} \subset X \times R$ be the open domain on which the flow is defined.

A Poincare map of γ is a mapping $P: W_0 \to W_1$ where:

1. $W_0, W_1 \subset S$ are open neighbourhoods of $m \in S$, and P is a C^k diffeomorphism ;
2. there is a function $\delta : W_0 \rightarrow R$ such that for all $x \in W_0$, $(x, \tau - \delta(x)) \in \mathcal{D}$ and $P(x) = F(x, \tau - \delta(x))$
3. if $t \in (0, \tau - \delta(x))$, then $F(x,t) \notin W_0$.

The idea of Poincare map is shown in Figure 6.1

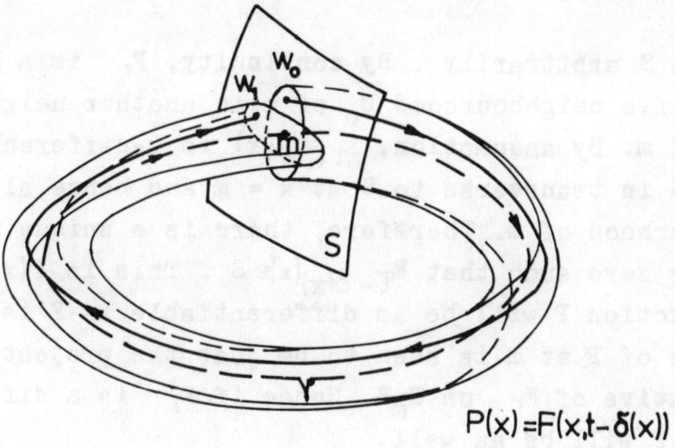

Figure 6.1

For the existence and uniqueness of Poincare maps we have the following theorem:

Theorem 6.1

1. If X is a C^k vector field on M, and γ is a closed orbit of X, then there exists a Poincare map of γ.

2. If $P: W_0 \to W_1$ is a Poincare map of γ (in local transversal section S at $m \in \gamma$) and P' also (in S' at $m' \in \gamma$), then P and P' are locally conjugated. That is, there are open neighbourhoods W_2 of $m \in S$, W_2' of $m' \in S'$, and a C^k diffeomorphism $H: W_2 \rightarrow W_2'$, such that $W_2 \subset W_0 \cap W_1$, $W_2' \subset W_0'$ and the diagram:

$$\begin{array}{ccc} \Theta^{-1}(W_2)\cap W_2 & \xrightarrow{\ \Theta\ } & W_2 \cap \Theta W_2 \\ \Big\downarrow H & & \Big\downarrow H \\ W_2' & \longrightarrow & S' \end{array}$$

commutes.

Proof sketch :

Choose S arbitrarily . By continuity, F_τ is a homeomorphism of a neighbourhood U_0 of m to another neighbourhood U_2 of m. By assumption, $F_{\tau+t}(x)$ is t-differentiable at t=0 and in transverse to S at x = m and hence also in a neighbourhood of m. Therefore, there is a unique number $\delta(x)$ near zero such that $F_{\tau-\delta(x)}(x)\in S$. This is P(x), and by construction P will be as differentiable as F is. The derivative of P at m is seen to be just the projection of the derivative of F_τ on T_mS. Hence if F_τ is a diffeomorphism, P will be as well.

First let's consider two particular examples of deterministic systems:

1. Let's take dynamics system described by the following autonomous equation:

$$\dot{\bar{x}} = \bar{f}(\bar{x}) \tag{6.1}$$

where $\bar{x} = [x_1,\ldots,x_n]^T$, $\bar{f} = [f_1,\ldots,f_n]^T$.
In this case if there exists the closed phase trajectory we can define Poincare map in the following way:

Definition 6.2

The set $M \subset R^2$:

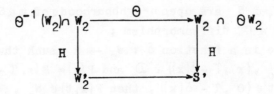

$$M \bar{x} t_o = \Big\{ (x_i(t),\ x_j(t)) \mid t = t_k,\ x_1(t_k) = \text{const}, \ldots,$$
$$x_{i-1}(t_k) = \text{const},\ x_{i+1}(t_k) = \text{const}, \ldots,$$
$$x_{j-1}(t_k) = \text{const},\ x_{j+1}(t_k) = \text{const}, \ldots,$$
$$x_n(t_k) = \text{const}\ ;\ k = 1,2,\ldots \Big\}$$

is a Poincare map for the system (6.1) if $n \geqslant 3$. $\bar{x}(t)$ is a solution of equation (6.1) with the initial condition $\bar{x}(t_o)$.

2. Let's take dynamics system descibed by nonautonomous equation :

$$\dot{\bar{x}} = \bar{f}(\bar{x}) + \bar{Q}(x) \qquad (6.2)$$

where $\bar{x} = [x_1, x_2]^T$, $\bar{f} = [f_1, f_2]^T$, $\bar{Q}(t) = [Q_1(t),\ Q_2(t)]^T$ $Q_i(t)$ $(i=1,2)$ are periodic functions of a period T. In this case we have a symetrical transposition:

$$S: (\bar{x}_i\ ,\ t) \longrightarrow (\bar{x}\ ,\ t+T)$$

and if there exists the closed phase trajectory γ we can define Poincare map in the following way:

Definition 6.3

The set $M \subset R^2$:

$$M(\bar{x}(t_o)) = \Big\{ (x_i(t),\ x_j(t)) \mid t = kT\ ,\ k=1,2,\ldots \Big\}$$

where x t is a solution of equation (6.2) with the initial

condition $\bar{x}(t_o)$, is a Poincare map of the system (6.2).

In the case of regular behaviour Poincare maps defined above consist of the finite numbers of points. In some cases when we deal with strange non-chaotic attractors these numbers can be large for example about 900 , (see Seydey (1986), Kapitaniak et al. (1987) , Romeiras and Ott (1987)). For chaotic behaviour the map consists of infinite number of points and has a structure of Cantor set:

$$\overline{\overline{\{M(\bar{x}(t_o))\}}} = \varkappa_o$$

where $\overline{\overline{\{\cdot\}}}$ indicates power of the set and $\varkappa_o$ is a power of natural numbers set.

In numerical calculations we have to restrict to finite numbers of points ($k = 10^3 - 10^4$ is most common). It is difficult to distinquish between strange chaotic and strange non-chaotic attractors based only on the finite approximations of the Poincare map , but these approximations are a good illustration of the attractor.

6.2. Mean Poincare maps

Definitions 6.1 - 6.3 describe Poincare map for deterministic systems. In the case of stochastic system (3.15) for each realization of random function $\eta(t,\omega)$, we obtain different Poincare map.

Now let's define mean Poincare maps:

Definition 6.4

Let's take the autonomous system perturbed by random noise:

$$\dot{\bar{x}} = \bar{f}(\bar{x}) + \bar{\eta}(t,\omega) \qquad (6.3)$$

The set $M \subset R^2$:

$$\langle M(\overline{x}(t_o))\rangle = \Big\{ (\langle x_i(t)\rangle, \langle x_j(t)\rangle) \mid t=t_k , \langle x_1(t_k)\rangle = const, \ldots, \langle x_{i-1}(t_k)\rangle = const , \langle x_{i+1}(t_k)\rangle = const, \ldots, \langle x_{j-1}(t_k)\rangle = const, \langle x_{j+1}(t_k)\rangle = const, \ldots, \langle x_n(t_k)\rangle = const, k=1,2,\ldots \Big\}$$

where $\overline{x}(t)$ are the realizations of the solution of equation (6.3) for initial condition $\overline{x}(t_o)$, $\langle\cdot\rangle$ indicates mean value, is called mean Poincare map.

In the same way for the system (6.2) perturbed by random noise we have the following definition of mean Poincare map:

Definition 6.5

Let's take the nonautonomous system perturbed by random noise:

$$\dot{\overline{x}} = \overline{f}(\overline{x}) + \overline{Q}(t) + \eta(t,\omega) \qquad (6.4)$$

The set $M \subset R^2$:

$$\langle M(x(t_o))\rangle = \Big\{ (\langle x_i(t)\rangle, \langle x_j(t)\rangle) \mid t = kT , k=1,2,\ldots \Big\}$$

where $\overline{x}(t)$ are the realizations of the solution of equation (6.4) for initial condition $\overline{x}(t_o)$, $\langle\cdot\rangle$ indicates mean value, T is a period of $Q(t)$, is called mean Poincare map.

The solutions of equations (6.3) and (6.4) are the stochastic processes $\bar{x}(t,\omega)$ given on the Cartezjan product of time interval $[0,\infty)$ and set of elementary events Ω, $\omega\in\Omega$. For settled time t_k $x_i(t_k,\omega)$ are random variables $x_i(\cdot,\omega)$: $\Omega \rightarrow R$, so their mean values exist, and also mean Poincare maps exist.

6.3. Mean Poincare maps of noisy Duffing equation

Consider non-linear oscillator:

$$\dot{x}_1 = x_2$$

$$\dot{x}_2 = -\frac{1}{25}x_2 + \frac{1}{5}x_1 - \frac{8}{15}x_1^3 + \frac{2}{5}\cos\varepsilon t + \sum_{k=1}^{N} A_k\cos(\nu_k t + \varphi_k) \qquad (6.5)$$

where ε, A_k, ν_k and φ_k are constant.

Quasiperiodic component of the equation (6.5) can be treated as an approximation of the band-limited white noise with a frequency interval $[\nu_{min}, \nu_{max}]$.

This equation in deterministic case has been investigated by Seydel (1986) and chaotic behaviour has been found for some particular values of ε.

The finite approximation (k=1000) of the Poincare map for $\varepsilon=0.04$ is shown in Figure 6.2(a). In Figure 6.2(b) and (c) the examples of the realizations of Poincare maps are shown (the variance of random noise D=0.05).

In Figure 6.2(d) the mean Poincare map of this system is shown. Mean map has been obtained on the basis of 200 realizations of the response $x(t,\omega)$. The following estimator of mean value has been used:

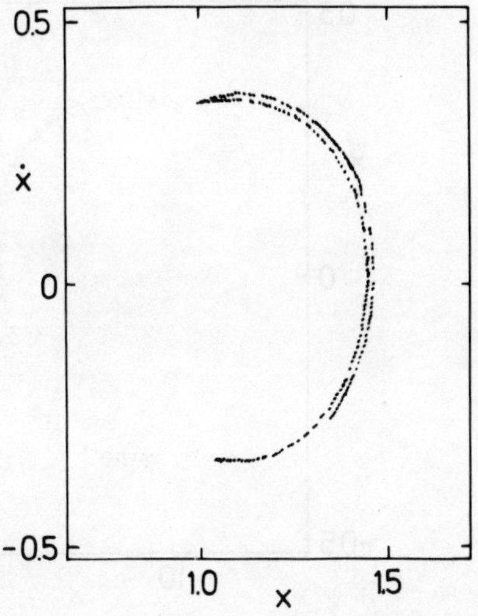

(a)

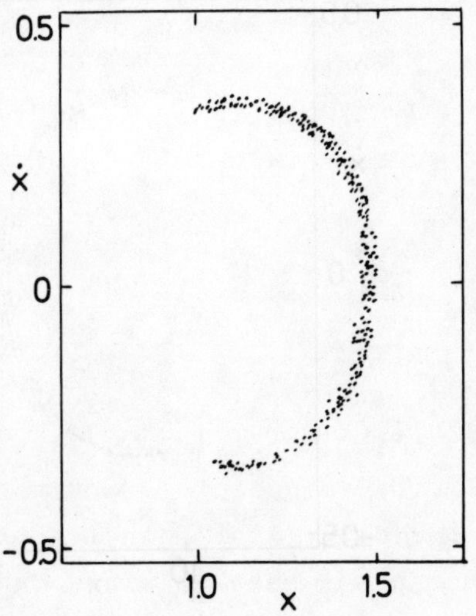

(b)

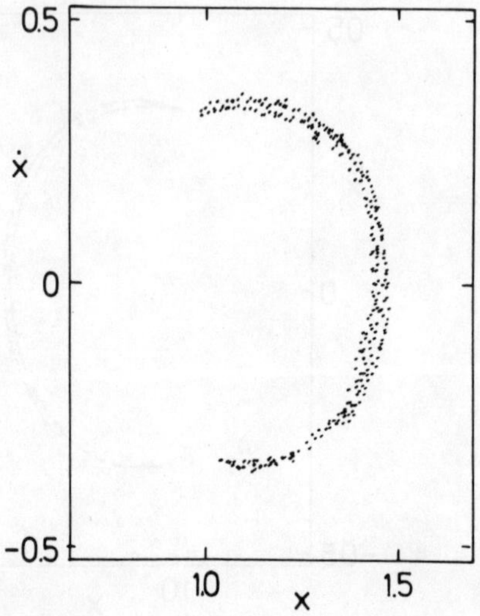

(c)

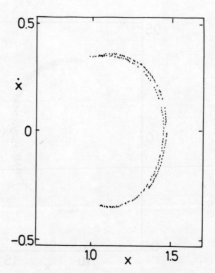

(d)

Figure 6.2

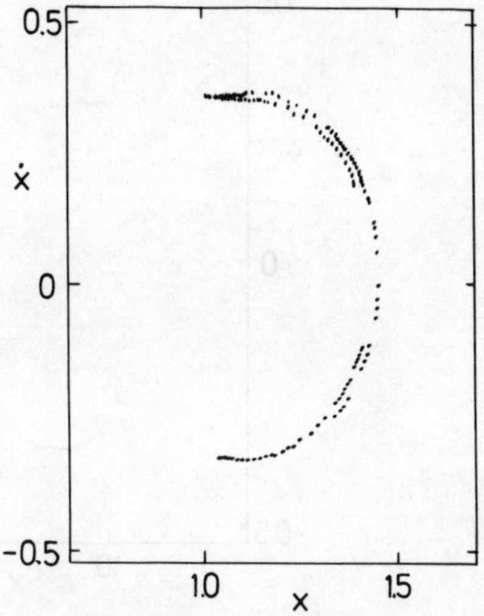

(a) D = 0.04

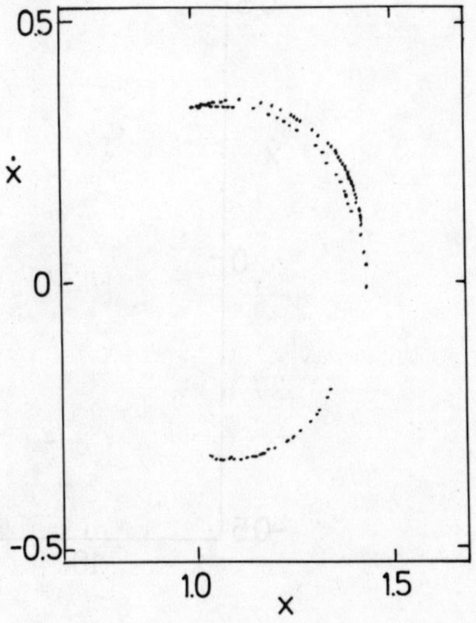

(b) D = 0.08

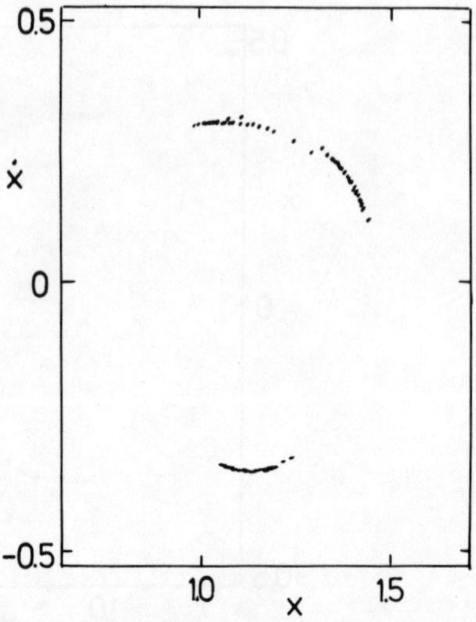

(c) D = 0.1

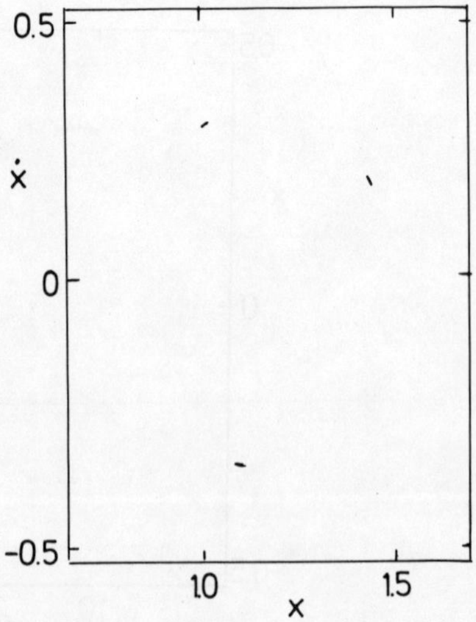

(d) D = 0.13

Figure 6.3: ε =0.04 , k=1000 , ν_{min}=0.01 , ν_{max}=0.1 .

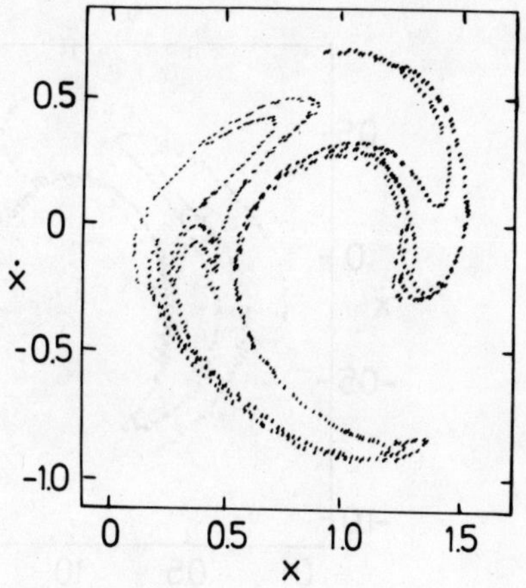

(a) D = 0.04

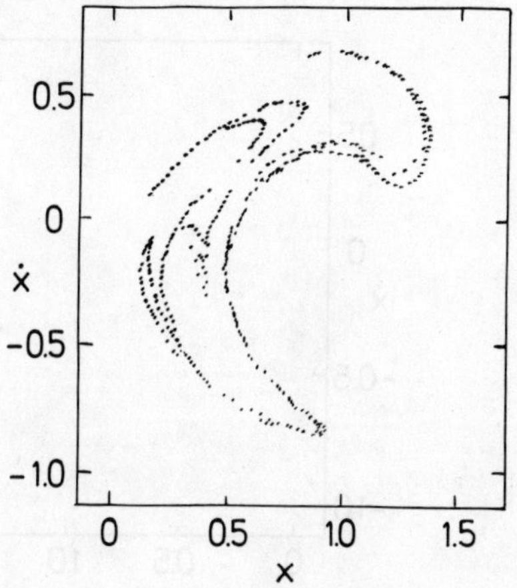

(b) D = 0.08

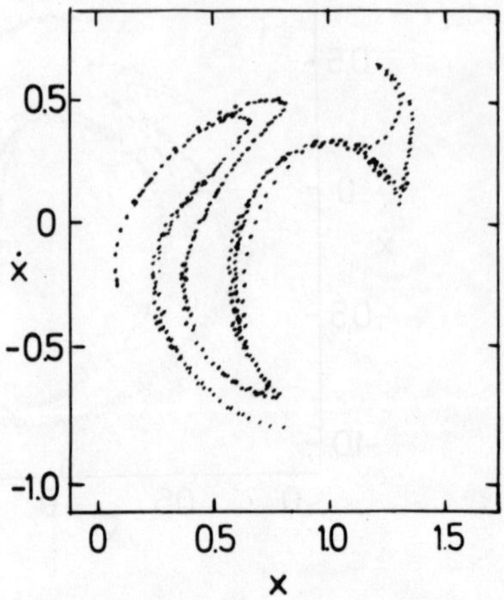

(c) D = 0.1

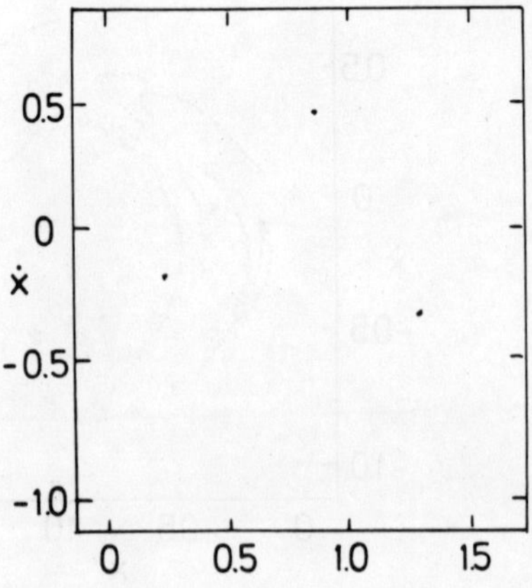

(d) D = 0.115

Figure 6.4 : $\varepsilon=0.2$, $k=1000$, $\nu_{min}=0.15$, $\nu_{max}=0.25$.

$$\langle \hat{x}(t) \rangle = \frac{1}{N} \sum_{n=1}^{N} x_n(t) \qquad t=kT \ , \ n=1,2,\ldots$$

The obtained in this case, structure of mean Poincare map is similar to the deterministic map shown in Figure 6.2(a).

Another example of mean Poincare maps for different values of noise variance D are shown in Figure 6.3(a)-(d).

With the increase of the noise variance D, up to D= 0.13 the structure of the mean maps shows the chaotic behaviour of the noisy system, but for larger values of D ($D \geqslant 0.13$) mean Poincare map consists of three points and the averaged attractor is characteristic for regular behaviour.

ε	$[\nu_{min}, \nu_{max}]$	N	c
0.04	[0.01 0.1]	30	0.130
		50	0.131
		70	0.130
	[0.02 0.12]	30	0.134
		50	0.135
		70	0.134
0.20	[0.15 0.25]	30	0.116
		50	0.116
		70	0.117
	[0.15 0.30]	30	0.147
		50	0.148
		70	0.148

Table 6.1

The same simplification of the mean Poincare maps has been observed for : $\varepsilon = 0.2$, $\nu_{min}=0.15$, $\nu_{max}=0.25$ - Figure 6.4(a)-(d). For the larger than 0.115 D the

the structure of the maps shows the regular behaviour of the system - Figure 6.4 (d)

The critical value of noise variance D for which chaotic system perturbed by quasiperiodic noise loses chaotic properties D_c , depends also on the interval of the noise frequencies ν_k, but seems to be independent of the number of harmonics - N To obtain a good approximation of band-limited white noise the number of harmonics must be relevantly big - (see Appendix).

Computed values of D_c for different frequencies intervals and different numbers of harmonics N are presented in Table 6.1

The analysis of the probability density function $P(x_1,t)$ of $x_1(t,\omega)$ shows that the mean Poincare map of the Cantor set structure :

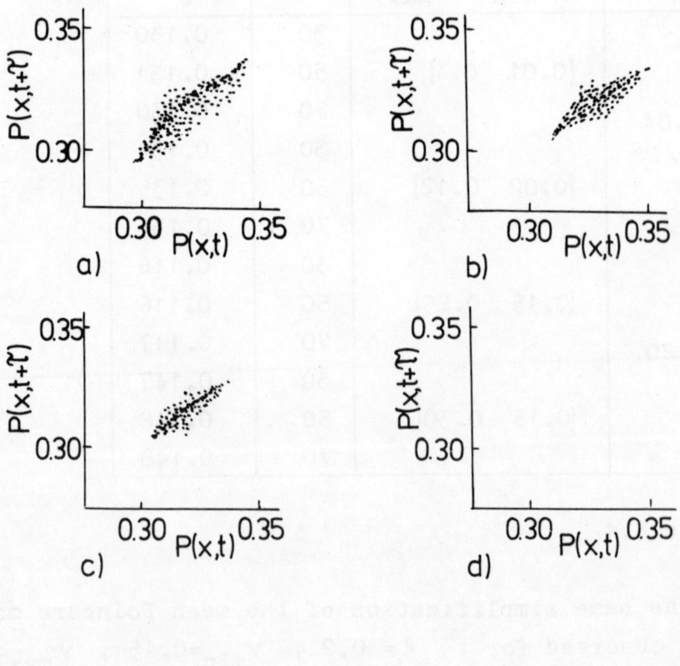

Figure 6.5

$$\overline{\overline{\left\{\langle M(\bar{x}(t_0))\rangle\right\}}} = \varkappa_0$$

is characteristic for chaotic stochastic processes. For example in Figure 6.5 (a) - (c) the maps $P(x_1,t) \rightarrow P(x_1, t+\tau)$ for $x_1=0$ and $\tau=2$ and parameters of Figure 6.3 (a) - (c) are shown. For comparison, the same map connected with Figure 6.3 (d) is shown in Figure 6.5 (d). In this case $P(x_1, t)$

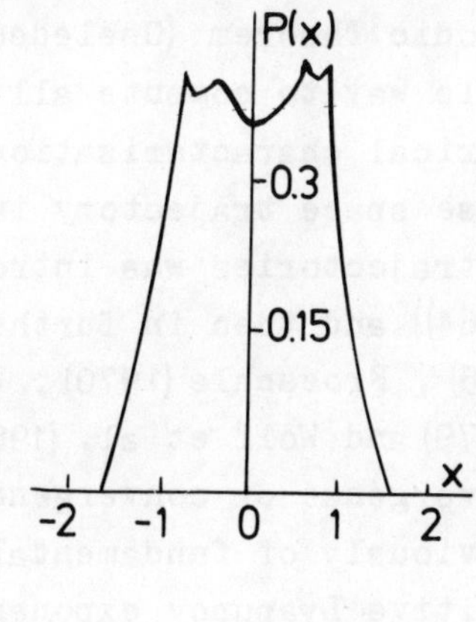

Figure 6.6

$\rightarrow P(x_1, t+\tau)$ map consists of one point and probability density function $P(x_1, t)$ is constant with its shape shown in Figure 6.6

7. RANDOM LYAPUNOV EXPONENTS

7.1. Deterministic Lyapunov exponents

Lyapunov exponents measure the mean exponential rate of divergence of nearby trajectories. The use of such exponents dates back to Lyapunov (Cesari (1959)). In a form adapted to the theory of dynamical systems and to ergodic theory they were introduced in 1968 when Oseledec published his non-communitative Ergodic Theorem (Oseledec (1968)) which provides general and simple way to compute all Lyapunov exponents. The first numerical characterization of the chaotic behaviour of the phase space trajectory in terms of the divergence of nearby trajectories was introduced in the work (Henon and Heiles (1964)) and then in further studies (Chirikov (1979), Ford (1975), Froeschle (1970), Wolf (1986), Shimada and Nagashima (1979) and Wolf et al. (1985)).

Rates of orbital divergence or convergence, called Lyapunov exponents are obviously of fundamental importance while studying chaos. Positive Lyapunov exponents indicate orbital divergence and chaos. Negative or equal to zero Lyapunov exponents are characteristic for regular behaviour.

One of the simplest examples of exponental divergence of nearby trajectories is the logistic map:

$$x(n + 1) = r\, x(n)(1 - x(n))$$

$x(0)$ is an initial condition chosen in the interval $(0, 1)$, r is tunable parameter in $[0, 4]$. For $r=4$ the sequence of map iterates $x(i)$, $(i=0,1,2,....)$ is known to be chaotic. In Figure 7.1 the trajectories from two nearby initial conditions are seen to be diverging already after three iterations.

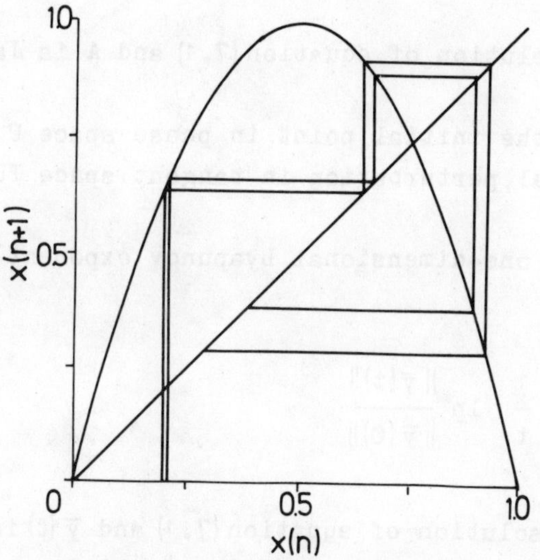

Figure 7.1

Now we give a definition of the largest maximum one -dimensional Lyapunov exponent λ_{max} for a system for which the equations of motion are explicitly known. For the method of computation of Lyapunov expoments from experimental data (see Wolf et al. (1985)).

Definition 7.1

Let's take dymamics system described by equation:

$$\dot{\overline{x}} = \overline{f}(\overline{x}) \tag{7.1}$$

where $\overline{x} = [x_1, \ldots, x_n]^T \in U$, U is an open set in R^n, $\overline{f} = [f_1, \ldots, f_n]^T$.

Let TU_x be the tangent space to U in the point $x \in U$. The tangent vector $y \in TU_x$ fulfils variational equation:

$$\dot{\bar{y}} = A(\bar{x}(t))\,\bar{y} \tag{7.2}$$

where x t is a solution of equation (7.1) and A is Jacobian $A(\bar{x}) = (\partial\bar{f}/\partial\bar{x})$.

Let $\bar{x}(0)$ be the initial point in phase space U and let $\bar{y}(0)$ be the initial perturbation in tangent space TU_{x_0} in the point $\bar{x}(0)$.

The maximum one-dimensional Lyapunov exponent is a limit:

$$\lambda_{max} = \lim_{t\to\infty} \frac{1}{t} \ln \frac{\|\bar{y}(t)\|}{\|\bar{y}(0)\|} \tag{7.3}$$

where $\bar{x}(t)$ is a solution of equation (7.1) and $\bar{y}(t)$ is a solution of variational equation (7.2), $\|\cdot\|$ indicates norm in R^n.

Now let's describe n-dimensional Lyapunov exponents. The solution of equation (7.2) can be written as

$$\bar{y}(t) = V^t_{x_0}\,\bar{y}(0)$$

where $V^t_{x_0}$ is the fundamental matrix of equation (7.2). The fundamental matrix of this equation satisfies the following chain rule:

$$V^{t+S}_{x_0} = V^t_{T\,x_0} \cdot V^s_{x_0}$$

where T is fundamental matrix of equation (7.1). It is apparent that the asymptotic behaviour of a small deviation is described by the asymptotic behaviour of the fundamental matrix for $t\to\infty$. Now, the asymptotic behaviour of this matrix for $t \to \infty$ can be characterized by the following exponents:

$$\lambda(e^k, x(0)) = \lim_{t\to\infty} \frac{1}{t} \ln \frac{\| V^t_{\bar{x}_o} \bar{e}_1 \wedge V^t_{\bar{x}_o} \bar{e}_2 \wedge \dots V^t_{\bar{x}_o} \wedge \bar{e}_k \|}{\| \bar{e}_1 \ \bar{e}_2 \ \dots \ \bar{e}_k \|}$$

for k=1,2,..,n . The symbols in the above formula have the following meanings: $\bar{e}^k$ is a k-dimensional subspace in the tangent space TU at $\bar{x}(0)$, $\{\bar{e}_i\}$ i=1,2,..,k are a set of bases of e^k, $\wedge$ is an exterior product and $\|\cdot\|$ is a norm in R^k. The exponent defined by the above formula represents an expanding rate of volume of the k-dimensional parallelepiped in the tangent space along the orbit which starts at x(0), and is called the k-dimensional Lyapunov exponent. It is clear from this definition that the exponent does not depend on a choice of set of bases or norms, but depends only on the k-dimensional subspace $\bar{e}^k$.

The main properties of the Lyapunov exponents are the following:

1. 1-dimensional exponent $\lambda(\bar{e}^1, \bar{x})$ may take, at most, n distinct values, and we have $\lambda_1 \geqslant \lambda_2 \geqslant \dots \lambda_n$.
2. k-dimensional exponent $\lambda(\bar{e}^k, \bar{x})$ may take at most, ${}_nC_k$ distinct values, and each value is connected with a sum of k distinct 1-dimensional exponents. For example for n=3 , the k-dimensional exponents $\lambda(\bar{e}^k, \bar{x})$ (k=1.2,3) may take the following values:

$\lambda(\bar{e}^1, \bar{x})$ = one of the values in $\{\lambda_1, \lambda_2, \lambda_3\}$

$\lambda(\bar{e}^2, \bar{x})$ = one of the values in $\{\lambda_1 + \lambda_2, \ \lambda_1 + \lambda_3, \ \lambda_2 + \lambda_3\}$

$\lambda(\bar{e}^3, \bar{x}) = (\lambda_1 + \lambda_2 + \lambda_3)$

3. If set of bases $\{\bar{e}_i\}$ (i=1,2,...,n) is chosen at random in tangent space, then the k-dimensional exponents $\lambda(\bar{e}^k,\bar{x})$ for k=1,2,..,n converge respectively, with probability 1, to the maximum values among sets of values which are allowed to possess ${}_nC_k$ distinct values.

It is obvious that Lyapunov exponents exist if the limit in 7.3 exists. We have the following theorem:

Theorem 7.1 the multiplicative ergidic theorem of Oseledec

If there is T^t - invariant measure μ and $\|\partial\bar{f}/\partial\bar{x}\| \in L^1(\mu)$, then the k-dimensional Lyapunov exponents $\lambda(\bar{e}^k,\bar{x}(0))$ k=1,2,..,n exist for μ -almost all $\bar{x}(0)$.

Now we have to remark some relations between Lyapunov exponents and the Kolmogorov entropy of dynamics systems. It has been known that the existence of the Lyapunov exponents is directly related to the Kolmogorov entropy. The weakest relation is as follows:

$$H(\mu) - \int \sum_{\lambda_i > 0} \lambda_i(\bar{x})\, d\mu \leqslant 0$$

where $H(\mu)$ is the Kolmogorov entropy of the dynamics system with invariant measure μ.

7.2. Lyapunov exponents for noisy systems

If the maximum Lyapunov exponent λ_{max} is computed in the presence of noise in the formula (7.2) the deterministic trajectory x will be replaced by the perturbed orbit (Arnold (1982), Herzel et al. (1987)). By this procedure λ_{max} is defined via linearization along noisy trajectory, and thus it describes the separation of nearby orbits subjected to the same external noise. To calculate

Lyapunov exponents in this way it is necessary to use the same realizations of the random noise for both orbits. Otherwise , we can expect in addition to an exponential separation described by $\lambda > 0$ a power low separation due to noise which is well-known from the phenomena of diffusion.

The examples of maximum one-dimensional Lyapunov exponent computed in this way are shown in Figure 7.2

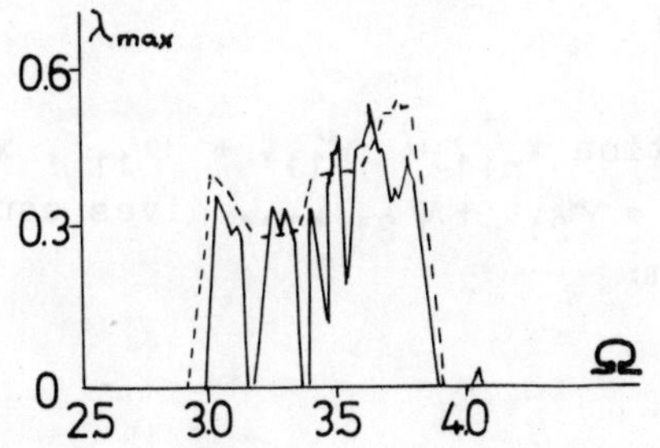

Figure 7.2

The full line shows the maximum Lyapunov exponent of the system:

$$\ddot{x} + \dot{x} - 10x + 100x^3 = \cos 3.5t + \eta(t,\omega) \qquad (7.4)$$

in the absence of white noise $\eta(t,\omega) = 0$, whereas the dashed line visualizes the smoothing effect of noise. It turns out that the thresholds of chaos are shifted and that the windows disappear.

Now let's describe maximum one-dimensional Lyapunov exponent for noisy system in the case when random noise $\bar{\eta}(t,\omega)$ is approximated by the sum of harmonics (see Appendix) $\bar{\eta}(t) = [\eta_1(t), \ldots, \eta_n(t)]^T$ where:

$$\eta_j(t) = \sum_{i=1}^{K} A_{ji} \cos(\nu_{ji} t + \varphi_{ji}) \tag{7.5}$$

φ_{ji}, $(i=1,2,\ldots,K\ ,\ j=1,2,\ldots,n)$ are the independent random variables with uniform distribution on interval $[0\ ,\ 2\pi]$ and A_{ji} and ν_{ji} can be computed in various ways depending on the method (see Appendix).

Making use of approximation (7.5) in equation (3.15) we obtain:

$$\dot{\bar{x}} = \bar{f}(\bar{x}) + \bar{\eta}(t) \tag{7.6}$$

Now the transposition $x_{n+1} = \nu_{11} t + \varphi_{11}$, $x_{n+2} = \nu_{12} t + \varphi_{12}$,...., $x_{n+K+1} = \nu_{21} t + \varphi_{21}$,... gives equation (7.6) in the following form:

$$\dot{\tilde{\bar{x}}} = \tilde{\bar{f}}\ \tilde{\bar{x}} \tag{7.7}$$

where:

$$\tilde{\bar{x}} = [x_1, \ldots, x_n, x_{n+1}, \ldots, x_{n+K}, x_{n+K+1}, \ldots, x_{n+nK}]^T$$

$$\tilde{\bar{f}} = [f_1 + \sum_{i=1}^{K} A_{1i} \cos x_{n+1}, \ldots, f_n + \sum_{i=1}^{K} A_{ni} \cos x_{n(1+i)},$$
$$\nu_{11}, \ldots, \nu_{nK}]^T$$

with equation (7.7) the following initial condition is connected:

$$\tilde{\bar{x}}(0) = [x_{10}, \ldots, x_{n0},\ \varphi_{11}, \ldots, \varphi_{nK}]^T$$

where $x_{10}, \ldots, x_{n0}$ are the initial conditions of equation (3.15).

The variational equation of equation (7.7) has the following form:

$$\dot{\tilde{\tilde{y}}} = A(\tilde{\tilde{x}}(t))\,\tilde{\tilde{y}} \tag{7.8}$$

with initial condition:

$$\tilde{\tilde{y}}(0) = [y_{10}, \ldots, y_{n0}, 0, \ldots, 0]^T$$

where $y_{10}, \ldots, y_{n0}$ are the initial conditions of variational equation (7.2).

Taking into account the above mentioned assumption the maximum Lyapunov exponent can be described as follows:

Definition 7.2

Maximum Lyapunov exponent for the random system (7.6) is the limit:

$$\lambda_{max} = \lim_{t \to \infty} \frac{1}{t} \ln \frac{\|\tilde{\tilde{y}}\ t\|}{\|\tilde{\tilde{y}}\ 0\|} \tag{7.9}$$

where $\tilde{\tilde{x}}(t)$ is a solution of equation (7.7) and $\tilde{\tilde{y}}(t)$ a solution of variational equation (7.8).

7.3. Quantifying chaos with random Lyapunov exponents

Defined in the last paragraph maximum Lyapunov exponent for random system depends on the parameters of approximation (7.5) A_{ji} and ν_{ji} which in some methods are random variables. The value of the Lyapunov exponent for single realization of these variables is not sufficient to state if the response of the system is a chaotic or a regular stochastic process.

Now the properties of distribution of random Lyapunov

exponents which allow to distinquish between regular and chaotic stochastic response of the system will be introduced.

Let's consider non-linear oscillator:

$$\ddot{x} + 0.1\dot{x} + x^3 = B\cos t + \eta(t,\omega) \tag{7.10}$$

$\eta(t,\omega)$ is band limited white noise with spectral density:

$$S_0(\nu) = \begin{cases} \dfrac{D}{\nu_{max} - \nu_{min}} & \nu \in [\nu_{min}, \nu_{max}] \\ 0 & \nu \notin [\nu_{min}, \nu_{max}] \end{cases} \tag{7.11}$$

D is variance of $\eta(t,\omega)$, $[\nu_{min}, \nu_{max}]$ an interval of considered frequencies , approximated as a sum of harmonics:

$$\eta(t) = \sum_{i=1}^{K} A_i \cos(\nu_i t + \varphi_i) \tag{7.12}$$

where φ_i are random variables of uniform distribution on interval $[0 , 2\pi]$, A_i and ν_i are given as follows (see Appendix):

$$A_i = \sqrt{2s(\nu)\Delta\mu}$$

$$\nu_i = (i + \tfrac{1}{2})\Delta\nu + \delta\nu_i + \nu_{min} \tag{7.13}$$

$$\Delta\nu = \frac{\nu_{max} - \nu_{min}}{K}$$

where $\delta\nu_i$ is uniformly distributed in the interval $[-0.1\Delta\nu,$

$0.1\Delta\nu]$.

Single realization of the process $\eta(t,\omega)$ is obtained by substituting drawed values of random variables φ_i and $\delta\nu_1$, computed values of A_i and ν_i to the formula (4.12).

After transposition : $x_1 = x$, $x_2 = x$, $x_3 = t$, $x_4 = \nu_1 t + \varphi_1$, ..., $x_{3+K} = \nu_K t + \varphi_K$ the equations (4.10) will be as follows:

$$\dot{x}_1 = x_2$$

$$\dot{x}_2 = -0.1x_2 - x_1^3 + B\cos x_3 + \sum_{i=1}^{K} A_i \cos x_{3+1}$$

$$\dot{x}_3 = 1 \qquad (7.14)$$

$$\dot{x}_4 = \nu_1$$

....

....

....

$$\dot{x}_{3+K} = \nu_K$$

where $x_3(0) = 0$, $x_4(0) = \varphi_1$,..., $x_{3+K}(0) = \varphi_K$.

The variational equations have the following form:

$$\dot{y}_1 = y_2$$

$$\dot{y}_2 = -0.1y_2 - 3x_1^2 y_1 + B[\sin x_3]y_3 + $$

$$+ \sum_{i=1}^{K} A_i[\sin x_{3+1}]y_{3+1}$$

$$\dot{y}_3 = 0 \qquad (7.15)$$

$$\dot{y}_4 = 0$$

......

......

$$\dot{y}_{3+K} = 0$$

Without loss of generality we can put : y_3, y_4 ,...., $y_{3+K} = 1$, which simplifies equations (7.15):

$$\dot{y}_1 = y_2$$

$$\dot{y}_2 = -0.1y_2 - 3x_1^2 y_1 + B\sin x_3 + \sum_{i=1}^{K} A_i \sin x_{3+1} \tag{7.16}$$

Maximum one-dimensional Lyapunov exponent can be calculated from the formula:

$$\lambda_{max}(x_1(0),\ x_2(0),\ 0\ ,\ \varphi_1, \ldots, \varphi_K\ ,\ y_1(0),\ y_2(0))$$

$$= \lim_{t\to\infty} \frac{1}{t} \ln \|\bar{y}(t)\|$$

where $x_1(0)$, $x_2(0)$ are the initial conditions of equation (7.14) and $y_1(0)$, $y_2(0)$ of the equation (7.16).

Maximum Lyapunov exponent is independent of the norm $\|\cdot\|$. In R^2 we have three standard norms:

$$\|\bar{y}\|_1 = \sum_{n=1}^{2} |y_n| \tag{7.17}$$

$$\|\bar{y}\|_2 = \sup_{1\leqslant n\leqslant 2} |y_n| \tag{7.18}$$

and Euclidian norm:

$$\|y\|_3 = \sqrt{\sum_{n=1}^{2} |y_n|^2} \tag{7.19}$$

In all calculations of Lyapunov exponents in this book the first norm (7.17) has been used. There are not any computational difficulties while using the second one (7.18) but the usage of Euclidian norm (7.19) may cause the overflows as y_1 and y_2 are large in the chaotic region.

For each realization of random process $\eta(t,\omega)$ the maximum Lyapunov exponent has been calculated. The two types of Lyapunov exponents distributions have been distinquished.

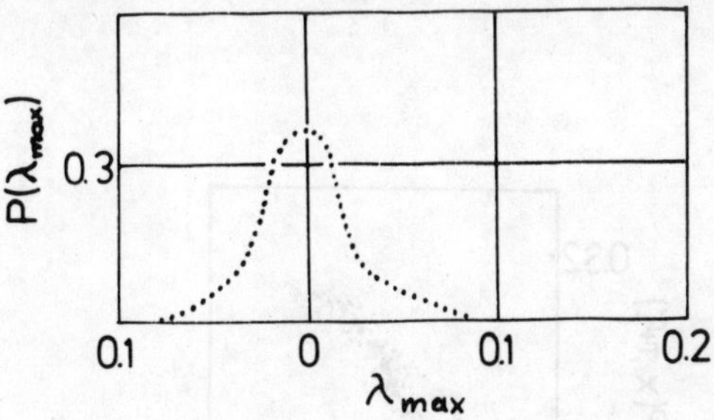

(a) B = 9.5

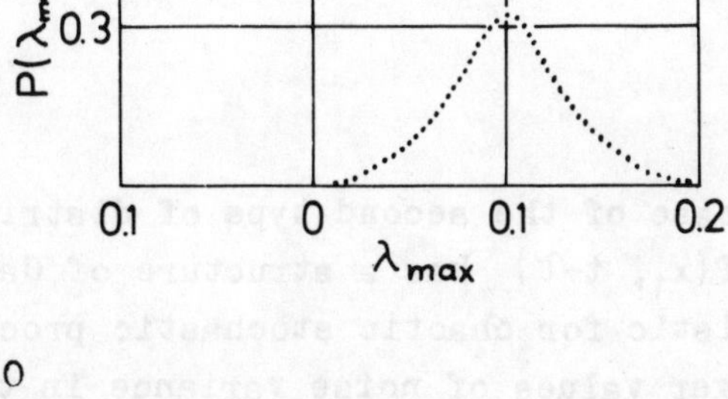

(b) B = 10.0

Figure 7.3: N = 30.0 , D = 0.02 .

The first one, where zero is the most probable value, is shown in Figure 7.3(a). The second distribution with positive mean value is shown in Figure 7.3(b). With the first type of distribution the $P(x_1, t) \rightarrow P(x_1, t+\tau)$ map characteristic for regular responses is connected - Figure

7.4 (a).

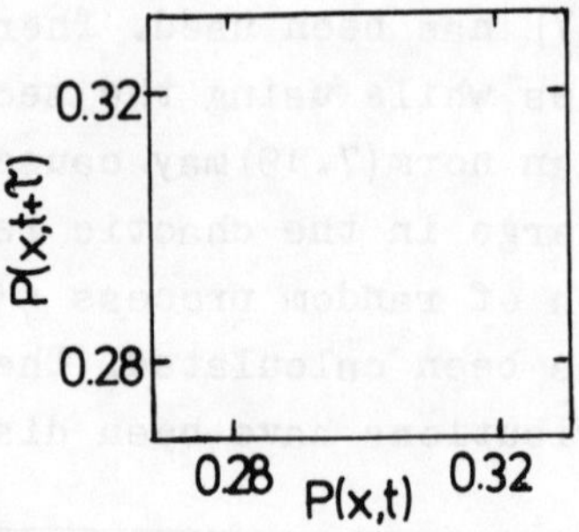

(a)

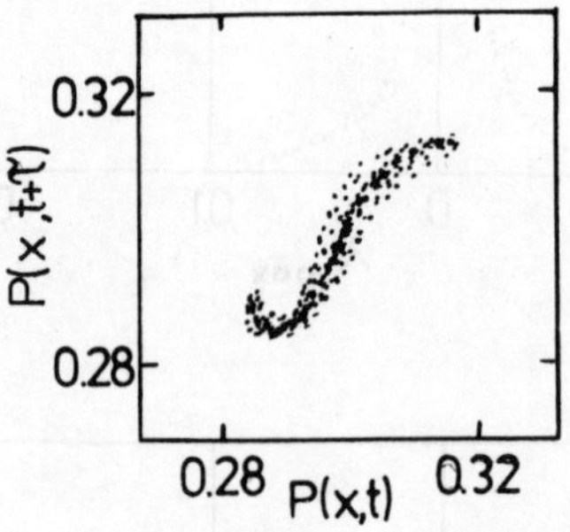

(b)

Figure 7.4

In the case of the second type of distribution , $P(x_1, t) \rightarrow P(x_1, t+\tau)$ has a structure of Cantor set and is characteristic for chaotic stochastic process.

For bigger values of noise variance in the system (7.10) the phenomenon of moving of the Lyapunov exponents distribution mean value toward zero has been observed - Figure 7.5 . Finally for the critical value of D which equals $D_c = 0.13$ B = 10.0 the distribution originally characteristic for chaotic response becomes characteristic for regular one .

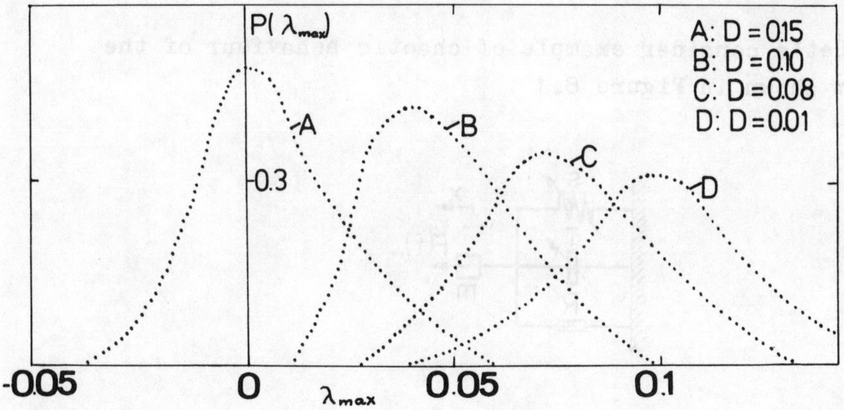

Figure 7.5 : B = 10.0

The system (7.10) shows chaotic behaviour if the amplitude of external excitation $B \in [9.9, 13.3]$. For

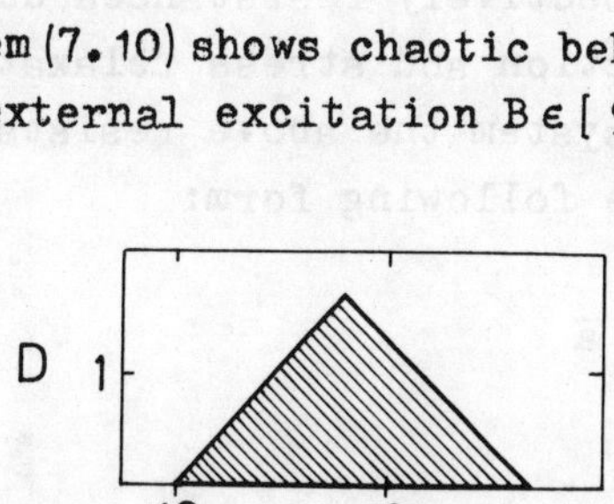

Figure 7.6

other values of B from this interval, we obtain different values of D_c. The plot D_c versus B has been shown in Figure 7.6 .

8. EXAMPLES

8.1. A simple mechanical system with stress relaxation

Let's consider example of chaotic behaviour of the system shown in Figure 8.1 .

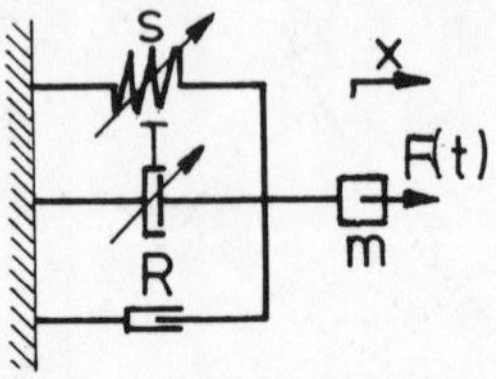

Figure 8.1

δ indicates internal force, x - displacement of the mass m , S , T and R are respectively resistances due to the stiffness, external friction and stress relaxation.

In the considered system the above resistances have been approximated in the following form:

$$S = (ex + dx^2 + fx^3)\ \bar{a} \tag{8.1}$$

$$T = (bx + cxx^2)\ \bar{a} \tag{8.2}$$

$$R = -\bar{a}\delta \tag{8.3}$$

that indicates non-linear nonsymmetric elastic characteristic , non-linear friction characteristic and linear stress relaxation characteristic.

Taking into consideration formulas (8.1 - 8.3) we obtain the following equations of motion of a body of mass m :

$$m\ddot{x} + \delta = F(t)$$
$$\dot{\delta} + a\delta = bx + cxx^2 + ex + dx^2 + fx^3 \qquad (8.4)$$

where $a = 1/\overline{a}$.

In the absence of the external excitation $(F(t) = 0)$, the self-excited oscillations of the system (8.4) occur.

System (8.4) in this case has one $(x_0 = \dot{x}_0 = \delta_0 = 0)$ or three singular points depending on the parameters a, b,...,f,m. The coordinates of nonzero singular points are:

$$x_{1,2} = \frac{d \pm \sqrt{d^2 - 4fe}}{2f} \qquad (8.5)$$

$$\dot{x}_{1,2} = 0 \qquad (8.6)$$

and

$$\delta_{1,2} = \frac{1}{d}[(b - a - 1)\ x_{1,2} + \frac{c}{3}x_{1,2}^3] \qquad (8.7)$$

According to the Routh-Hurwitz criterion, the conditions for non-stability of n-th singular point (n=0,1,2) are:

- for the aperiodic instability

$$[2c - 3f]x_n^2 - 2x_n - e > 0 \qquad (8.8)$$

- for the oscillating instability

$$[ca - 3f]x_n^2 + 2dx_n + ba - e < 0 \qquad (8.9)$$

We study the behaviour of the solutions of the system (8.4) in the instability regions of all singular points. For some values of the system parameters the solution changes from periodic or almost periodic to chaotic one. The sets of parameters for which the system shows chaotic behaviour have been estimated by calculating maximum one-dimensional Lyapunov exponent and chaotic behaviour has been found for the parameters a, c, d from the zone shown in Figure 8.2 and parameters b, f and e fulfilling the following relations:

$$b = \alpha - a$$

$$f = \frac{c}{3} + \beta d \qquad (8.10)$$

$$e = \alpha - \gamma d$$

where $\alpha \in [2.2\ ,\ 3.9]$, $\beta \in [0.65\ ,\ 0.9]$ and $\gamma \in [0.9\ ,\ 1.2]$.

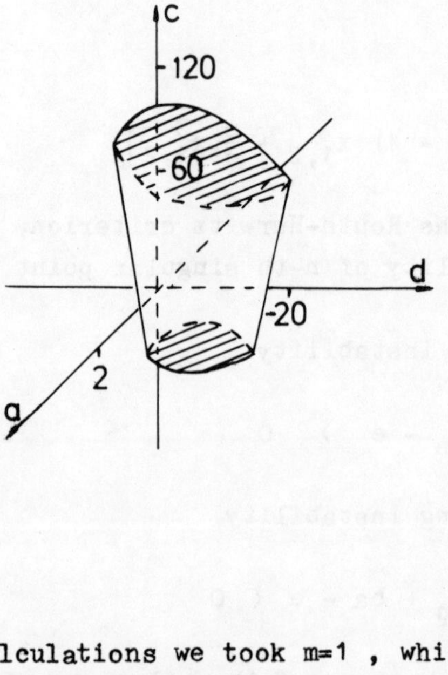

Figure 8.2

In our calculations we took m=1 , which does not decrease the generality of them.

The examples of Poincare maps of the plane $\dot{x}=0.8$ and $\dot{x}=1.2$ have been shown in Figure 8.3 .

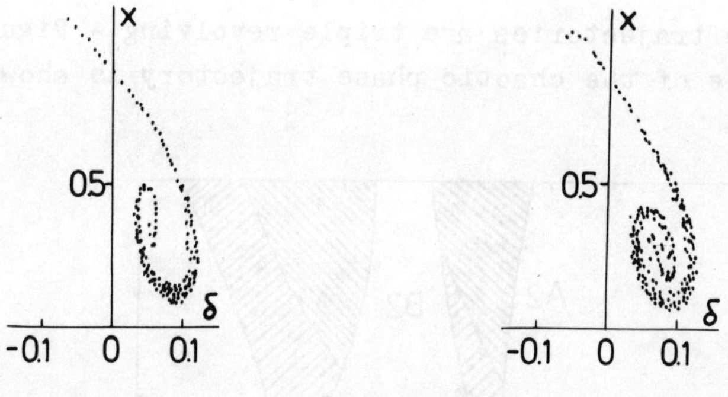

(a) $\dot{x} = 0.8$ (b) $\dot{x} = 1.2$

Figure 8.3

In the chaotic zone all singular points are unstable and are of saddle-focus type.

Now consider the complete system (8.4) with the following external force:

$$F(t) = A \cos \Omega_0 t + \eta(t,\omega) \tag{8.11}$$

where A and Ω_0 are constant and $\eta(t,\omega)$ is band-limited white noise stochastic process approximated by formula (7.12) with coefficients (7.13).

First the system with only periodic deterministic excitation has been considered. For the parameters values, for which an autonomous system shows chaotic behaviour and for the particular values of A and Ω_0 we find the zone where the system is not chaotic - Figure 8.4 . In the zone A the solution is periodic with a period $4\pi/\Omega_0$, and in zone B periodic with a period $2\pi/\Omega_0$. In the zones A and B there are subzones A1 , B1 where the phase trajectories are double revolving - Figure 8.5 and the zones A2 , B2 where the phase trajectories are triple revolving - Figure 8.6 . An example of the chaotic phase trajectory is shown in Figure 8.7

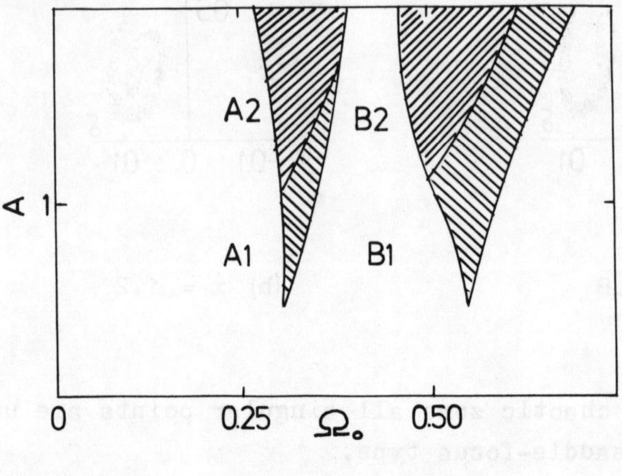

Figure 8.4 : a=1.75 , b=1.9 , c=54.2 , d=-20.0 , e=21.2 , f=2.1 , m=1.0 .

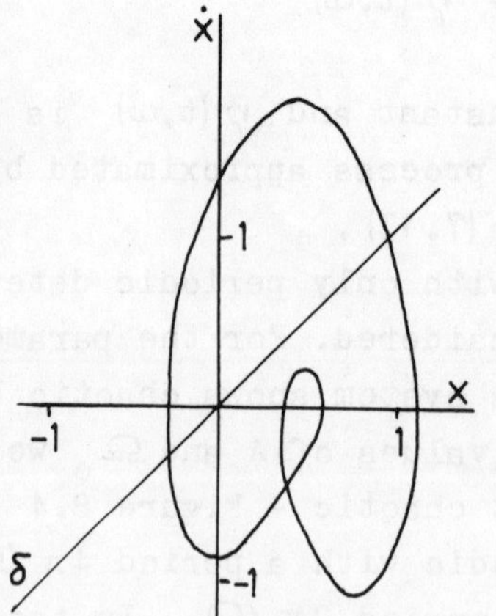

Figure 8.5 : A=1.5 , Ω_0=0.6 .

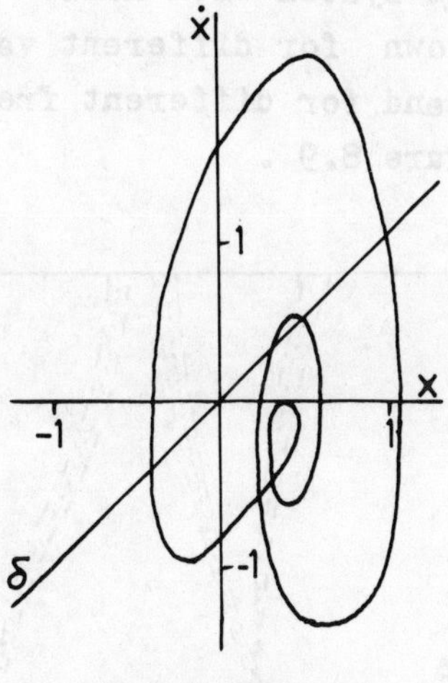

Figure 8.6 : A=1.5 , Ω_o=0.5 .

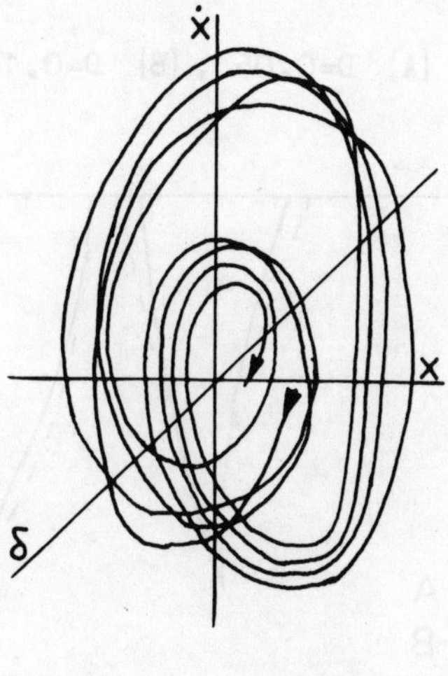

Figure 8.7 : A=1.5 , Ω_o=0.32 .

In the Figure 8.8 and 8.9 the zones of the chaotic behaviour of the system with noise chaotic stochastic response are shown for different values of noise variance - Figure 8.8 and for different frequencies intervals $[\nu_{min}, \nu_{max}]$ - Figure 8.9 .

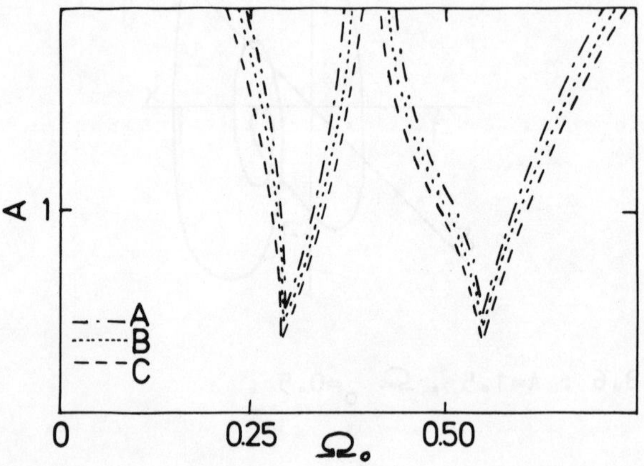

Figure 8.8 : (A) D=0.05 , (B) D=0.1 , (C) D=0.2

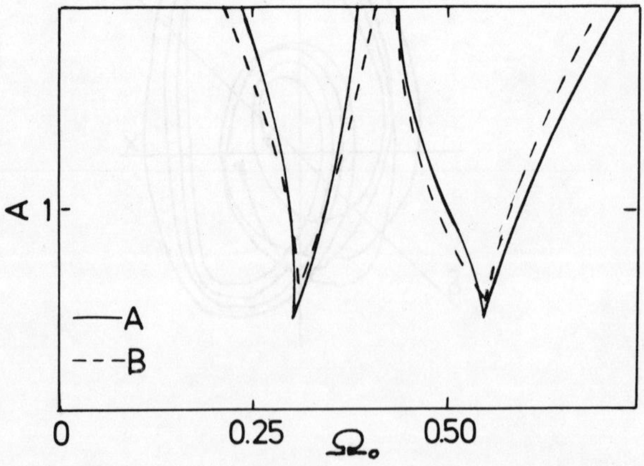

Figure 8.9 : D=0.1 , (A) $\nu \epsilon [0.4 , 1.8]$, (B) $\nu \epsilon [0.8 , 1.4]$

As in the deterministic case in the zones A and B the behaviour of the system is randomly perturbed regular one. It is interesting that in the case of additional random perturbations these zones increase with the increase of the noise variance.

8.2. Non-linear oscillator with quasi-periodic excitation

As the next example consider non-linear oscillator (7.10) assuming that the amplitude of the external excitation is as follows:

$$B = 2f(t,\omega)\cos\,\varepsilon t \tag{8.12}$$

where $f(t,\omega)$ is a random function with constant mean B and variance D .

Taking into account formula (8.12) the equation (7.10) will have the following form:

$$x + 0.1x + x^3 = 2f(t,\omega)\cos\varepsilon t\cos t \tag{8.13}$$

$$= f(t,\omega)\left[\cos(1-\varepsilon)t + \cos(1+\varepsilon)\,t\right]$$

The above equation is the particular example of the equation forced by two external periodic forces which will be considered in the Chapter 10.

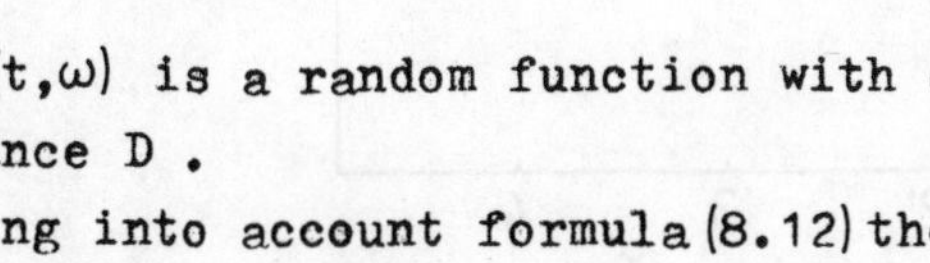

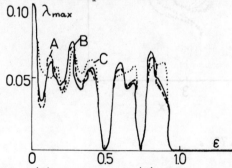

Figure 8.10 : (A) B = 10.0 , (B) B = 11.5 , (C) B = 13.0 .

In deterministic case D=0 this equation has been investigated (Kapitaniak et al. 1987) and chaotic behaviour has been found for $B \in [4.95 , 6.65]$ and $\varepsilon \in [0 , 0.95] - \{0.5 , 0.75\}$. The examples of maximum Lyapunov exponents are shown in Figure 8.10 .

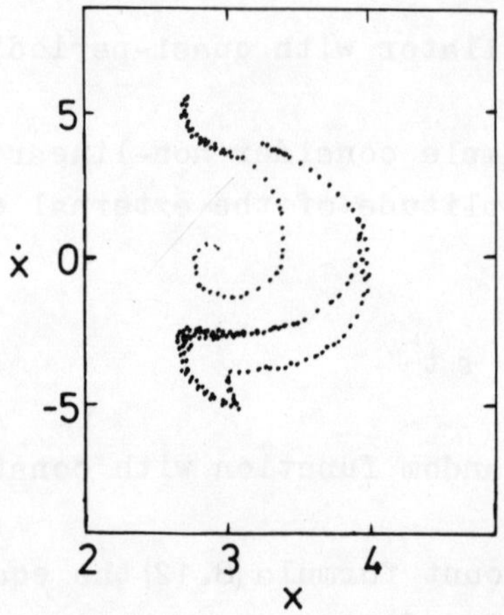

(a) B=10.0 , ε =0.5 .

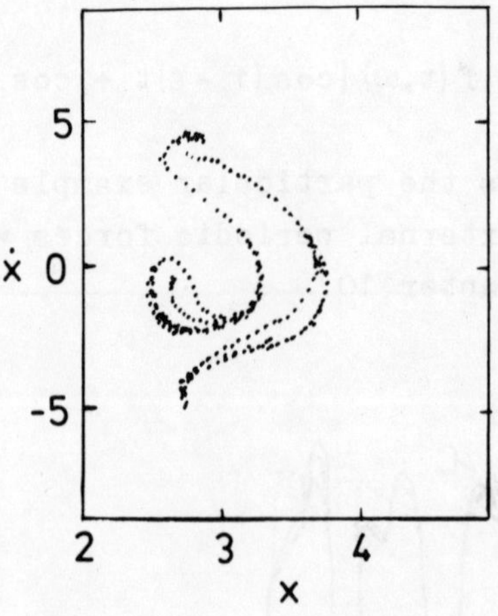

(b) B=10.0 , ε =0.75 .

Figure 8.11

In the interval of values of the parameter ε, for which the system (8.13) shows chaotic behaviour, two isolated points (0.5 and 0.75) for which the system shows quasi-periodic behaviour, have been determined. The attractor in these points is very complicated , as shown in Poincare maps presented in Figure 8.11 these are the examples of the so-called strange non-chaotic attractors.

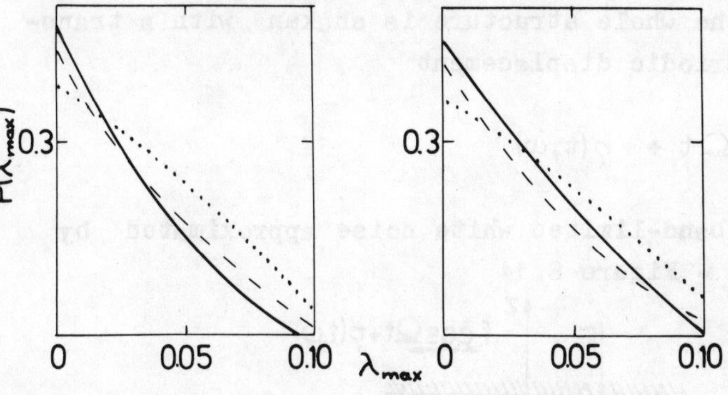

(a) ε =0.5 (b) ε =0.75

Figure 8.12 : ____ D=0.1 , - - - D=0.15 , ... D=0.2 .

The aim of this example is to investigate what kind of behaviour the noisy system in these points presents.

In Figure 8.12 the distributions of Lyapunov exponents for ε =0.5 , 0.75 and D = 0.2 , 0.4 , 0.6 are shown. In all cases, the type of distribution is characteristic for regular stochastic process.

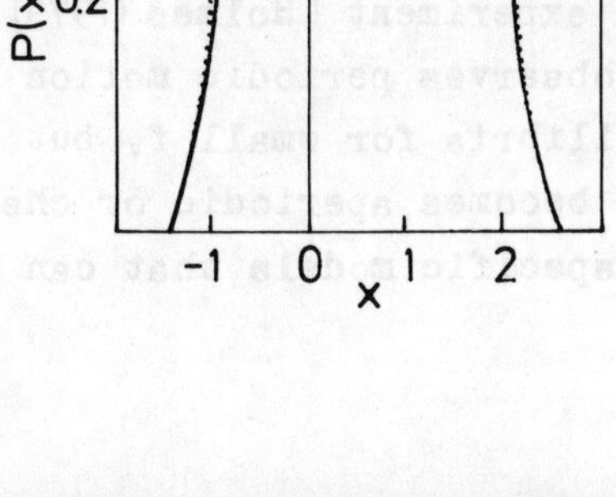

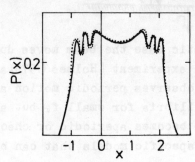

Figure 8.13

The same is found on the basis of the probability density function of $x(t,\omega)$ which is constant in time and has the shape shown in Figure 8.13 .

8.3. Chaotic oscillations of the buckled beam

Consider a beam that is buckled by an external load , so there are two stable and one unstable equilibrium states. Then the whole structure is shaken with a transverse noisy periodic displacement

$$f\cos \Omega t + \eta(t,\omega)$$

$\eta(t,\omega)$ is a band-limited white noise approximated by formula (7.12) - Figure 8.14

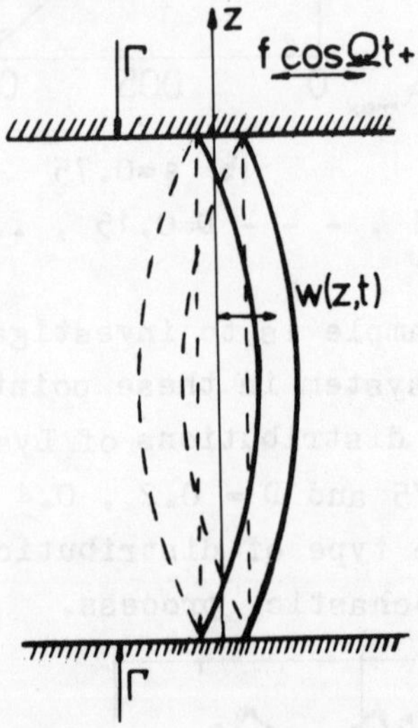

Figure 8.14

In the deterministic case the beam moves due to its inertia. In a related experiment (Holmes (1979) and Moon and Holmes (1981)) one observes periodic motion about one of the two stable equilibria for small f, but as f is increased, the motion becomes aperiodic or chaotic. There is a number of specific models that can be used to describe the beam in Figure 8.14 . One of them is the following partial differential equation for the transverse deflection w(z, t) of the center line of the beam:

$$\ddot{w} + w'''' + \Gamma w'' - \varkappa\left(\int_0^1 [w']^2 d\xi\right) w'' \tag{8.14}$$

$$= f\cos\Omega t - a\dot{w} + \eta(t,\omega)$$

where $\dot{} = \partial/\partial t$, $' = \partial/\partial z$, Γ - external load, $\varkappa$ stiffness due to "membrane" effects, a - damping . Amongst many possible boundary conditions we shall choose $w=w''=0$ at $z=0, 1$; i.e. simply supported or hinged ends. With these boundary conditions the eigenvalues of the linearized unforced equations, i.e., complex numbers λ such that:

$$\lambda^2 w + w'''' + \Gamma w'' = 0 \tag{8.15}$$

for some non-zero w satisfying $w=w''=0$ at $z=0, 1$ form a countable set

$$\lambda_j = \pm \pi_j \sqrt{\Gamma - \pi^2 j^2} , \quad j=1, 2,...$$

Thus if $\Gamma \ \pi^2$ all eigenvalues are imaginary and the trivial solution $w=0$ is formally stable for positive damping it is Lyapunov stable . We shall henceforth assume that

$$\pi^2 < \Gamma < 4\pi^2$$

in which case the solution $w = 0$ is unstable with one positive and one negative eigenvalue and the nonlinear equation (8.14) with $a, f=0$ has two nontrivial stable buckled equilibrium states.

A simplified model for the dynamics of (8.14) is obtained by seeking the lowest mode solutions of the form

$$w\ (z,\ t) = x(t)\ \sin(\pi z) \tag{8.16}$$

Substituting it into (8.14) and taking the inner product with the basic function $\sin \pi z$ gives us a Duffing type of equation for the model displacement $x(t)$:

$$x + \quad ax - \quad bx + \quad cx^3 \quad = \quad \gamma \cos \Omega t + \ \eta(t,\omega) \tag{8.17}$$

where $b = \pi^2(\Gamma - \pi^2) > 0$, $c = x\pi^4/2$, $\gamma = 4f/\pi$.

The equation (8.17) has the same form as the equation (7.4) and in the deterministic case shows chaotic behaviour for $a=0.1$, $b=10.0$, $c=100.0$, $\gamma=1$ and $\Omega \in [3.0\ ,\ 4.1]$. Maximum one-dimensional Lyapunov exponent for different values of Ω has been shown in Figure 7.2 full line .

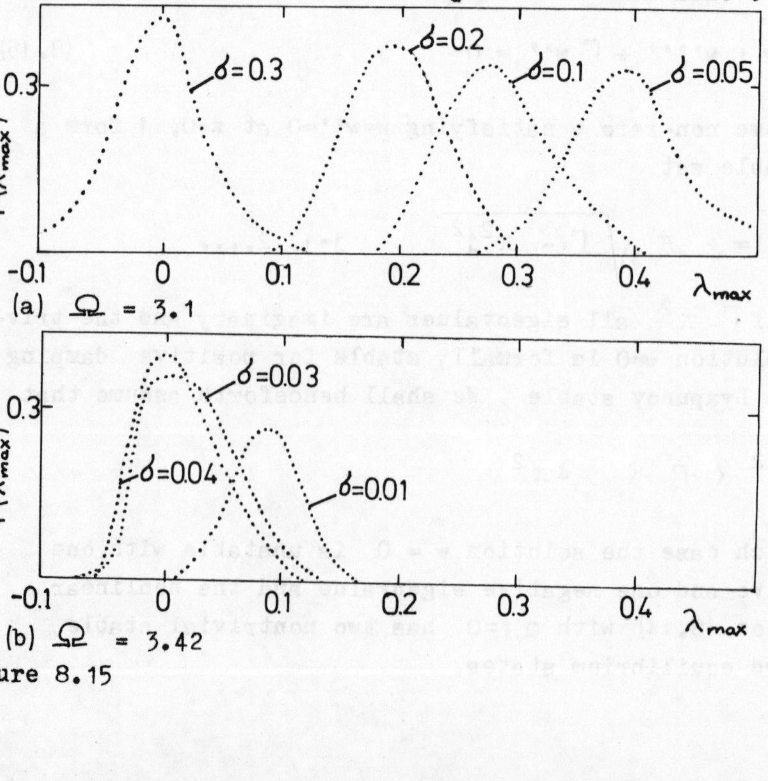

(a) Ω = 3.1

(b) Ω = 3.42

Figure 8.15

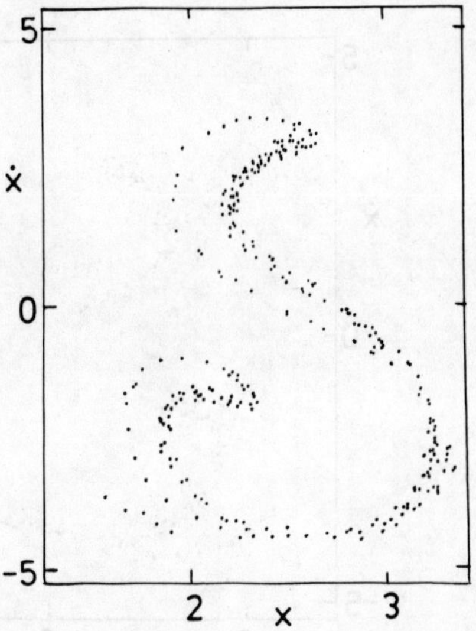

(a) Ω = 3.1 , δ = 4D/π = 0.1

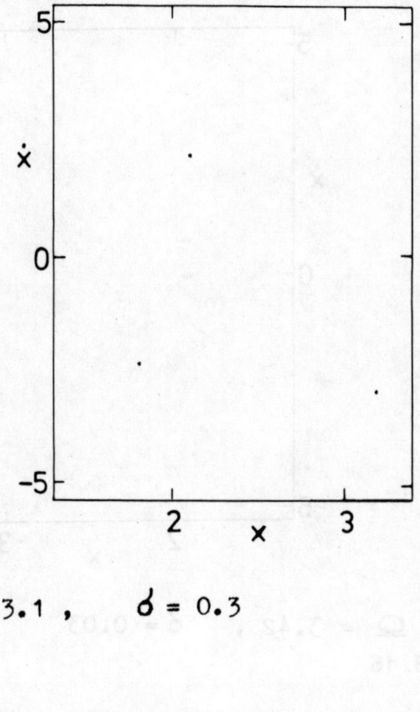

(b) Ω = 3.1 , δ = 0.3

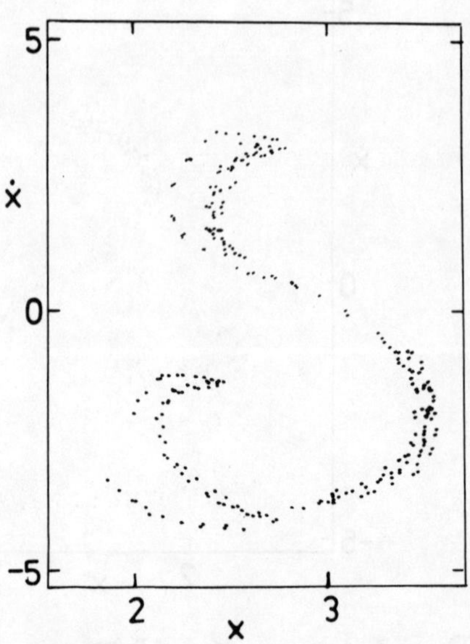

(c) Ω = 3.42 , δ = 0.02

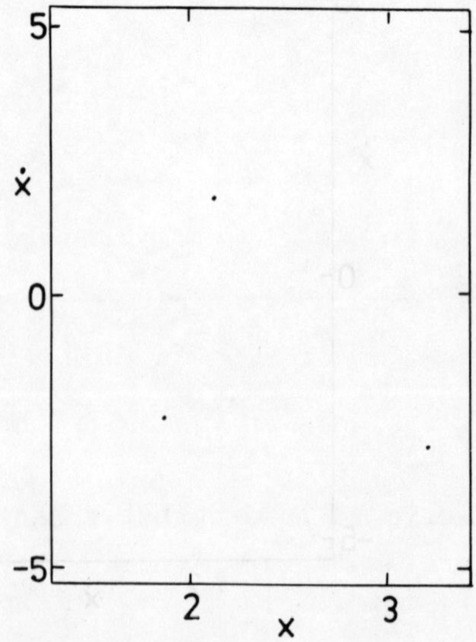

(d) Ω = 3.42 , δ = 0.03

Figure 8.16

In Figure 8.15 (a) - (b) the distributions of random maximum Lyapunov exponents for different Ω and noise variance σ have been shown. The examples of equivalent mean Poincare maps have been shown in Figure 8.16 (a) - (d). In the case of positive mean value of Lyapunov exponents distribution we obtain mean Poincare map characteristic for chaotic stochastic process.

In this example we observe the same phenomena as in the paragraph 7.3 of moving of the Lyapunov exponents distribution mean value towards zero with the increase of the noise variance.

It is interesting that for smaller values of deterministic λ_{max} the critical value of noise variance (the value for which the distribution becomes characteristic for regular stochastic process) is smaller than in the case of larger λ_{max}.

8.4. Noisy chaotic movement of a simple mechanical machine

Let's consider the coupled system of rotor and pendulum shown in Figure 8.17 (Zgone and Grabec (1987)).

The equations of motion can be written in the following form:

$$I_1 \ddot{\varphi}_1 = - M_{fr} + M_{mot}$$
$$I_2 \ddot{\varphi}_2 = - mgl\sin\varphi_2 + M_{fr} \qquad (8.18)$$

where φ_1 and φ_2 are angular variables and I_1 and I_2 are the moments of inertia.

In this example we assume that the driving moment falls down linearly when rotor angular velocity approches to a certain value w_m:

$$M_{mot}(\dot{\varphi}_1) = D\,(w_m - \dot{\varphi}_1 + \eta(t,\omega)) = B - D\,\dot{\varphi}_1 + \eta(t,\omega)$$

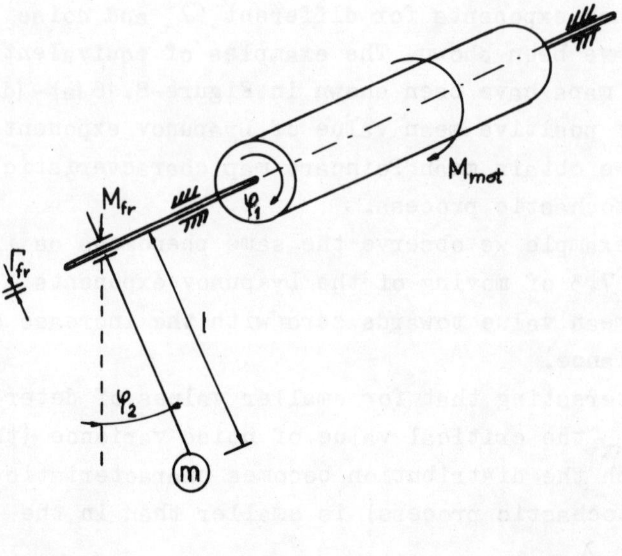

Figure 8.17

where $B = Dw_m$ denotes the starting moment and D the slope of the characteristic curve shown in Figure 8.18, $\eta(t,\omega)$ is the random force which perturbes the driving moment,

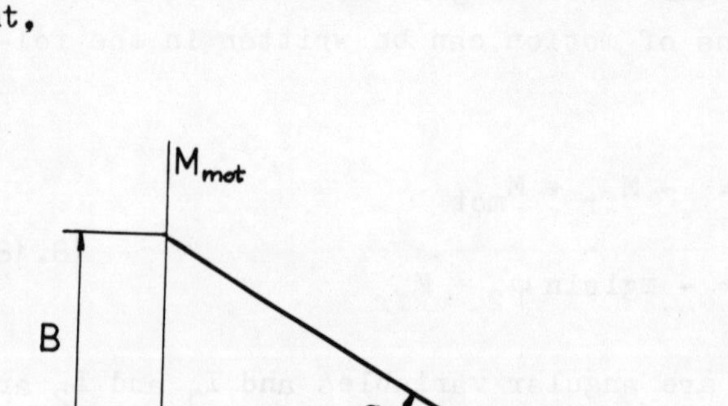

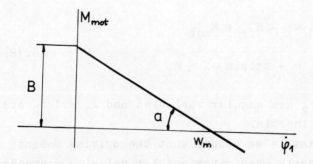

Figure 8.18

The effect of velocity on the frictional moment can be quite generally represented by a non-linear characteristic curve in Figure 8.19 which can be described by the following formula:

$$M_{fr}(w) = M_0[C(|\frac{w}{w_0}| - 1)^2 + 1]\,\mathrm{sgn}|w| \qquad (8.20)$$

where $w = \dot{\varphi}_1 - \dot{\varphi}_2$ is the sliding velocity, sgn denotes the sigmum function. The constants M_0 , C and w_0 depend on the properties of the construction of the system.

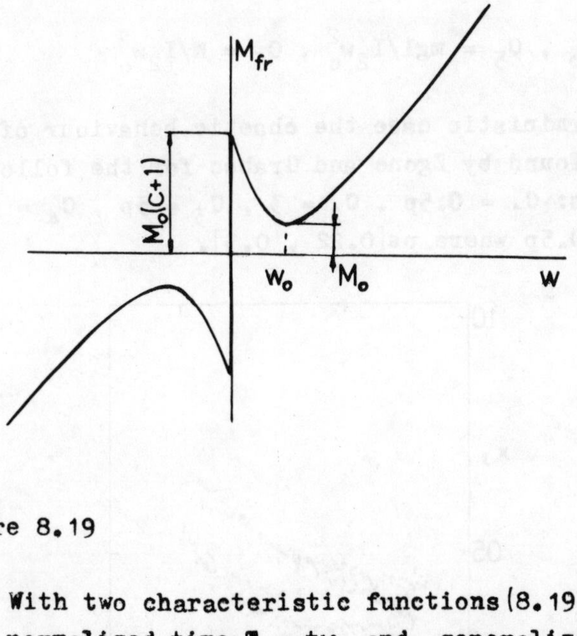

Figure 8.19

With two characteristic functions (8.19) and (8.20), with normalized time $T = tw_0$ and generalized coordinates

$$x_0 = \varphi_1 \; , \quad x_1 = \dot{\varphi}_1 \; , \quad x_2 = \varphi_2 \; , \quad x_3 = \dot{\varphi}_2$$

the equations (8.18) take the following form:

$$\dot{x}_0 = x_1$$

$$\dot{x}_1 = -C_1 f + C_3 + C_4 x_1 + \eta(T,\omega)$$

$$\dot{x}_2 = x_3 \qquad (8.21)$$

$$\dot{x}_3 = -C_5 \sin x_2 + C_6 f$$

where:

$$f = C_2\ (|x_1 - x_3| - 1)^2\ \mathrm{sgn}|x_1 - x_3|$$

$$C_1 = M_0/I_1 w_0^2\ ,\ C_2 = C\ ,\ C_3 = B/I_1 w_0^2$$

$$C_4 = D/I_1 w_0\ ,\ C_5 = mgl/I_2 w_0^2\ ,\ C_6 = M/I_2 w_0^2$$

In the deterministic case the chaotic behaviour of the system has been found by Zgone and Grabec for the following parameters values: $C_1 = 0.5p$, $C_2 = 3$, $C_3 = 5p$, $C_4 = 10p$, $C_5 = 0.5$, $C_6 = 0.5p$ where $p \in [0.22\ ,\ 0.3]$.

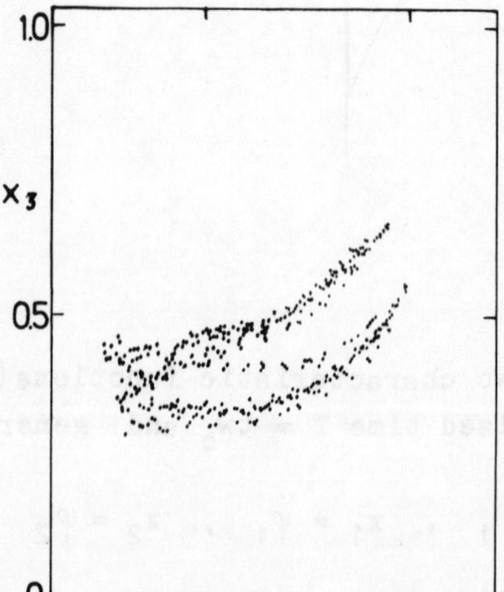

(a) D = 0

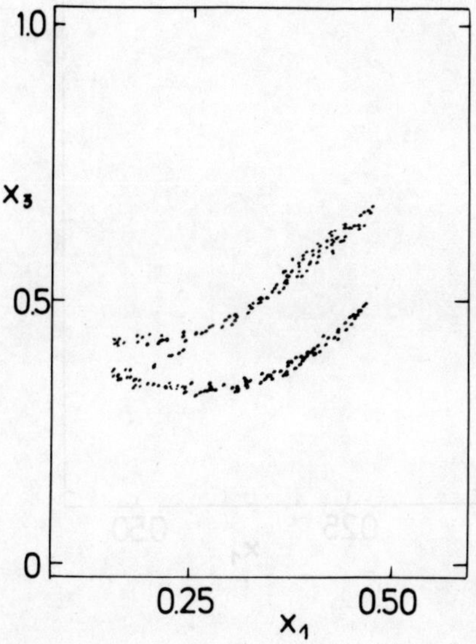

(b) D = 0.05

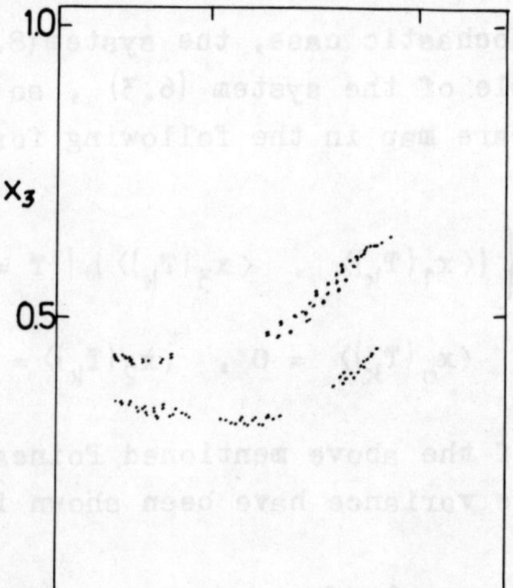

(c) D = 0.2

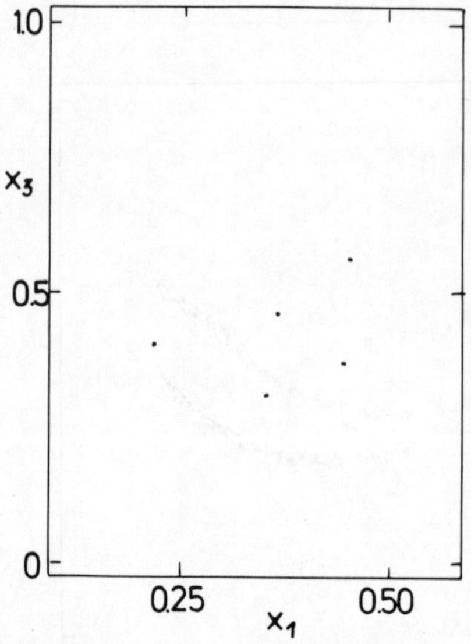

(d) D = 0.5

Figure 8.20 : p = 0.25

In the stochastic case, the system (8.21) is the specific example of the system (6.3), so we can define the mean Poincare map in the following form:

$$\langle M(x(T_0))\rangle = \Big\{ (\langle x_1(T_k)\rangle , \langle x_3(T_k)\rangle) \mid T = T_k , \langle x_0(T_k)\rangle = 0 , \langle x_2(T_k)\rangle = 0 \Big\}$$

The examples of the above mentioned Poincare maps for different noise variance have been shown in Figure 8.20 (a)-(d).

As in the example from paragraph 6.3 the structure of mean Poincare map has become simpler with the increase of noise variance.

9. STOCHASTIC SENSITIVITY FUNCTIONS AND CHAOS

The theory of sensitivity is very useful in the investigations of the influence of changes of parameters on the solution of dynamics system. The description of this theory in the stochastic case can be found in Szopa (1985).

Now following Szopa and Bestle (1986) we describe the property of stochastic sensitivity function which gives information on whether the stochastic system is chaotic or not.

Going back to the equation (2.1):

$$\dot{\bar{x}} = \bar{f}(\bar{x}, t, \omega) \tag{9.1}$$

we define the stochastic sensitivity function which describes the sensitivity of the system to its initial condition.

Definition 9.1

Let's assume that there exist the partial derivatives of the function $\bar{f}$ with respect to $\bar{x}$, and that the initial conditions are deterministic:

$$\bar{x}(0) = \bar{\xi}_0 = [\xi_{10}, \xi_{20}, \ldots, \xi_{n0}]^T$$

Let's assume that there exists the solution of equation (9.1):

$$\bar{x} = \bar{x}(\bar{\xi}_0, t, \omega) \tag{9.2}$$

The derivative with respect to $\bar{\xi}$ of $\bar{x}(\bar{\xi}_0, t, \omega)$ is called the stochastic sensitivity function:

$$\bar{S}(t,\omega) = [\,\sigma_{ij}(t,\omega)\,] = \left.\frac{\partial \bar{x}(\bar{\mathfrak{z}}, t, \omega)}{\partial \bar{\mathfrak{z}}}\right|_{\mathfrak{z} = \mathfrak{z}_o} \qquad (9.3)$$

$(i, j = 1, 2, \ldots, n)$.

Taking into account the derivative with regard to of equation (9.1) we obtain differential equations for the stochastic sensitivity function $S(t,\omega)$

$$\dot{\bar{S}}(t,\omega) = \left.\frac{\partial \bar{f}(\bar{x}, t, \omega)}{\partial \bar{x}}\right|_{\bar{x} = \bar{x}(\bar{\mathfrak{z}}_o, t, \omega)} \bar{S}(t,\omega) \qquad (9.4)$$

The initial conditions for equation (9.4) are given by the identity matrix:

$$\bar{S}(0,\omega) = \bar{I}$$

The property of stochastic sensitivity function of the chaotic system can be demostrated by considering Duffing's oscillator:

$$\ddot{y}(t,\omega) + 0.05\dot{y}(t,\omega) + y^3(t,\omega) = A(\omega)\cos t \qquad (9.5)$$

where $A(\omega)$ is a random variable uniformly distributed in the interval [7.4 , 7.6]. The initial conditions are as follows: $\bar{\mathfrak{z}}_o = [\,3.5\ ,\ 0\,]^T$.

In this case we obtain the following equation for the stochastic sensitivity function:

$$\dot{\bar{S}}(t,\omega) = \begin{bmatrix} 0 & 1 \\ -3y^2(t,\omega) & -k \end{bmatrix} \bar{S}(t,\omega) \qquad (9.6)$$

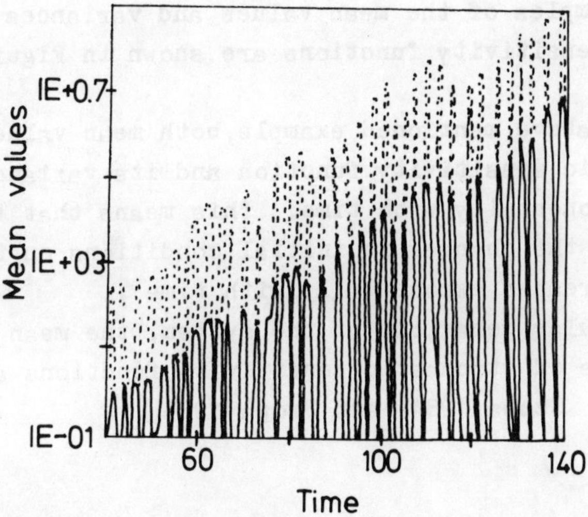

Figure 9.1 : Mean values of δ_{11} .... and δ_{12} —— logarithmic scale .

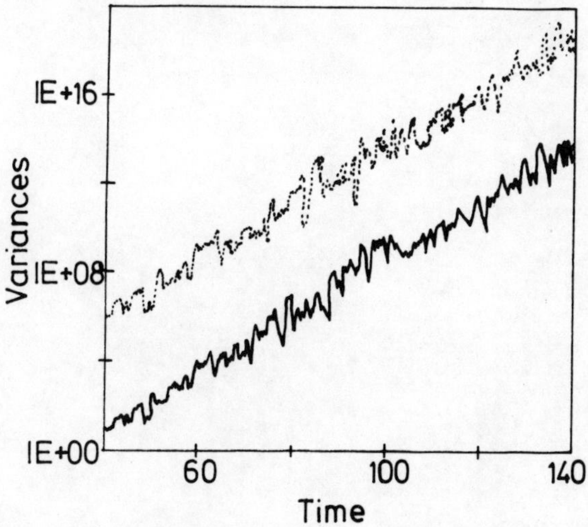

Figure 9.2 : Variances of δ_{11} and δ_{12} logarithmic scale .

where $\overline{x} = [\, y \,,\, \dot{y}\,]^T$.

The examples of the mean values and variances of stochastic sensitivity functions are shown in Figure 9.1 and 9.2 .

In the above mentioned example, both mean values of the stochastic sensitivity function and its variances increase exponentally with time. This means that the influence of the changes in initial conditions on the solution increases exponentally with time .

For regular behaviour of the system, the mean values and variances of stochastic sensitivity functions are limited (see Szopa (1984) and Szopa (1985)).

10. NOISY ROUTES TO CHAOS

The mechanism of the transition to chaos is of fundamental importance for understanding the phenomenon of chaos. It has been investigated recently by many authors for example : Szemplińska-Stupnicka and Bajkowski (1986), Szemplińka-Stupnicka (1986), Kapitaniak (1987), Novak and Frehlich (1982), Coullet and Tresser (1984), Argoul and Arneodo (1984). The most often route to chaos leads via the cascade of period doubling bifurcations Coullet and Tresser (1984), Novak and Frehlich (1982), but also other routes are possible, for example sharp transition : Szemplińska-Stupnicka (1986), Kapitaniak (1987).

10.1. The effect of noise on the logistic equation

First, let's consider a classical example of chaotic behaviour i.e. logistic equation

$$x[n+1] = r\,x[n](1 - x[n]) \qquad 0 < x_n < 1$$

which has also been mentioned in Chapter 7 . For r=4 the sequence of map iterates x[i], (i=0,...,∞) is known to be chaotic. The illustration of this map has been shown in Figure 7.1.

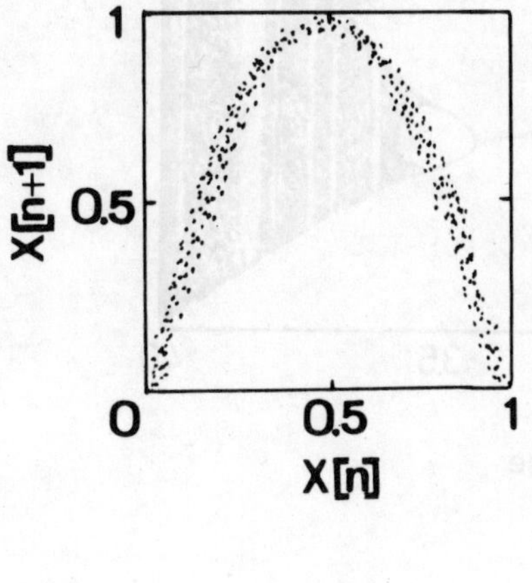

Figure 10.1

In Figure 10.1 we have the illustration of noisy logistic map:

$$x[n+1] = (r + q[n])\ x[n]\,(1 - x[n])$$

where q[n] indicates the fluctuations of the r parameter (Gaussian process with variance D=0.001 has been taken).

The bifurcation diagrams x[n] versus r in deterministic and noisy cases have been shown in Figures 10.2(a)-(b).

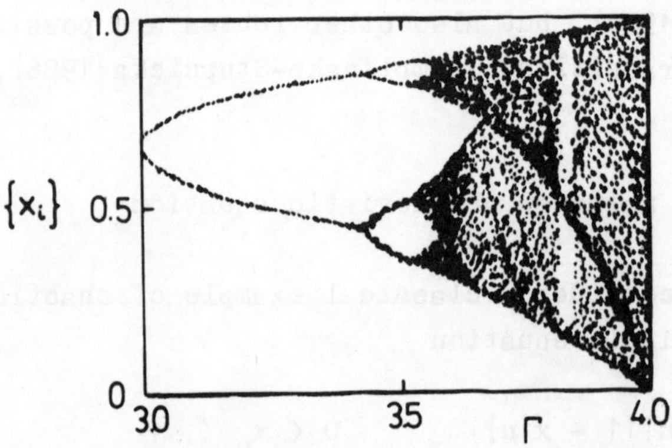

(a) deterministic case

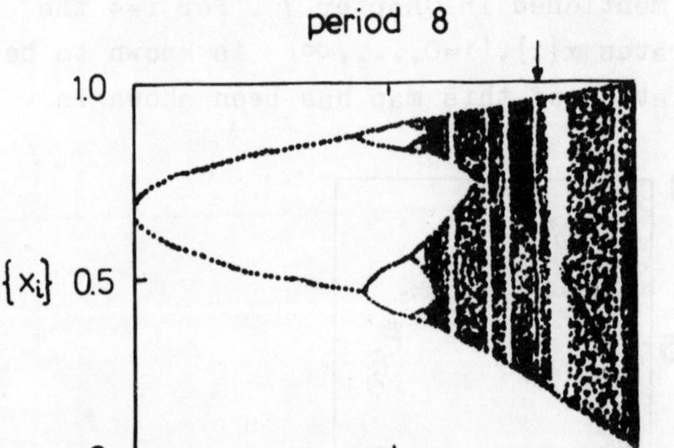

(b) stochastic case

Figure 10.2

Some structures which are characteristic for deterministic case are still visible in the presence of noise. For example, it is evident from Figure 10.2(b) that in the chaotic regime a window with a period three orbit is still present , although windows of higher periodicity are not visible. Of course with the increase of noise the numbers of attractors whose periods cannot be observed at a given noise level increased (see Crutchfield et al. (1982)).

10.2. Noisy route to chaos in non-linear oscillator with quasi-periodic excitation

Let's consider a non-linear oscillator:

$$x + ax + x^3 = F_1 \cos \Omega_1 t + F_2 \cos \Omega_2 t + \eta(t,\omega) \qquad (10.1)$$

where Ω_1 and Ω_2 are independent from each other, and $\eta(t,\omega)$ is white noise with variance D.

For $\Omega_2 = 0$ and $D=0$ the system (10.1) leads to the oscillator excited by one periodic and one constant external forces which has been investigated in Szemplińska-Stupnicka and Bajkowski (1986). In case of $\Omega_1 = \Omega - \varepsilon$, $\Omega_2 = \Omega + \varepsilon$, $F_1 = F_2 = F/2$ and $D=0$ we obtain the equation (8.13).

10.2.1 Deterministic case

First, let's consider the route to chaos in deterministic case ($\eta(t,\omega) = 0$).

In the first approximation, we consider the solution of equation (10.1) consisting of two components with frequencies Ω_1 and Ω_2 :

$$x(t) = C_1 \cos(\Omega_1 t + \nu_1) + C_2 \cos(\Omega_2 t + \nu_2) \qquad (10.2)$$

Substituting equation (10.2) into equation (10.1) and equating coefficients of $\cos(\Omega_1 t + \nu_1)$, $\sin(\Omega_1 t + \nu_2)$, $\cos(\Omega_2 t + \nu_2)$ and $\sin(\Omega_2 t + \nu_2)$ separately to zero, a set of algebraic equations for C_1, C_2, ν_1 and ν_2 has been obtained:

$$\begin{aligned} & - C_1\Omega_1^2 + \tfrac{3}{4}C_1^3 + \tfrac{3}{2}C_1C_2^2 = F_1\cos\nu_1 \\ & - aC_1\Omega_1 = F_1\sin\nu_1 \\ & - C_2\Omega_2^2 + \tfrac{3}{4}C_2^3 + \tfrac{3}{2}C_1^2C_2 = F_2\cos\nu_2 \\ & - aC_2\Omega_2 = F_2\sin\nu_2 \end{aligned} \tag{10.3}$$

In order to determine the first order correction to x we give:

$$\begin{aligned} x(t) = {} & C_1\cos(\Omega_1 t + \nu_1) + C_2\cos(\Omega_2 t + \nu_2) \\ & + C_3\cos(3\Omega_1 t + \nu_3) + C_4\cos(3\Omega_2 t + \nu_4) \end{aligned} \tag{10.4}$$

Neglecting the terms involving higher-order harmonics and keeping only those terms linear in C_3 and C_4, we determine the amplitudes C_3, C_4 and phases ν_3, ν_4 to be:

$$\begin{aligned} & - 9C_3\Omega_1^2 + \tfrac{3}{4}C_3^3 + \tfrac{3}{2}C_3C_4^2 = \tfrac{1}{4}C_1^3\cos(\nu_1 - \nu_3) \\ & - 3aC_3\Omega_1 = \tfrac{1}{4}C_1^3\sin(\nu_1 - \nu_3) \\ & - 9C_4\Omega_2^2 + \tfrac{3}{4}C_4^3 + \tfrac{3}{2}C_3^2C_4 = \tfrac{1}{4}C_2^3\cos(\nu_2 - \nu_4) \\ & - 3aC_4\Omega_2 = \tfrac{1}{4}C_2^3\sin(\nu_2 - \nu_4) \end{aligned} \tag{10.5}$$

To examine the local stability of the solution, (10.4) has been perturbed by σ, where $\sigma \ll x$:

$$\tilde{x}(t) = x(t) + \delta(t) \tag{10.6}$$

After inserting (10.6) into equation (10.1) and taking into account equations (10.3) and (10.5) we obtain the following variational equation:

$$\ddot{\delta}(t) + a\dot{\delta}(t) + 3x^2(t)\,\delta(t) + 3x(t)\,\delta^2(t) + \delta^3(t) = 0 \tag{10.7}$$

The local stability is examined neglecting non-linear variational equation. Then inserting equation (10.4) we have:

$$\begin{aligned}
&\ddot{\delta}(t) + a\dot{\delta}(t) + \delta(t)[\lambda_0 + \lambda_1\cos 2\Phi_1 + \lambda_2\sin 2\Phi_1 \\
&+ \lambda_3\cos 2\Phi_2 + \lambda_4\sin 2\Phi_2 + \lambda_5\cos 2\Phi_3 + \lambda_6\cos 2\Phi_4 \\
&+ \lambda_7\cos 4\Phi_1 + \lambda_8\sin 4\Phi_1 + \lambda_9\cos 4\Phi_2 + \lambda_{10}\sin 4\Phi_2 \\
&+ \lambda_{11}\cos(\Phi_1 - \Phi_2) + \lambda_{12}\cos(\Phi_1 - \Phi_2) + \lambda_{13}\cos(\Phi_1 + \Phi_4) \\
&+ \lambda_{14}\cos(\Phi_1 - \Phi_4) + \lambda_{15}\cos(\Phi_2 + \Phi_3) + \lambda_{16}\cos(\Phi_2 - \Phi_3) \\
&+ \lambda_{17}\cos(\Phi_3 + \Phi_4) + \lambda_{18}\cos(\Phi_3 - \Phi_4)] = 0
\end{aligned} \tag{10.8}$$

where:

$$\begin{aligned}
\lambda_0 &= \tfrac{1}{2}(C_1^2 + C_2^2 + C_3^2 + C_4^2) \\
\lambda_1 &= \tfrac{1}{2}C_1^2 + C_1C_3\cos(\nu_3 - 2\nu_1) \\
\lambda_2 &= -C_1C_3\sin(\nu_3 - 2\nu_1) \\
\lambda_3 &= \tfrac{1}{2}C_2^2 + C_2C_4\cos(\nu_4 - 2\nu_2)
\end{aligned} \tag{10.9}$$

$\lambda_4 = -C_2C_4\sin(\nu_4 - 2\nu_2)$

$\lambda_5 = \frac{1}{2}C_2^2$, $\lambda_6 = \frac{1}{2}C_4^2$, $\lambda_7 = C_1C_3\cos(\nu_3 - 2\nu_1)$

$\lambda_8 = \lambda_2$, $\lambda_9 = C_2C_4\cos(\nu_4 - 2\nu_2)$, $\lambda_{10} = \lambda_4$

$\lambda_{11} = \lambda_{12} = C_1C_2$, $\lambda_{13} = \lambda_{14} = C_1C_4$

$\lambda_{15} = \lambda_{16} = C_2C_3$, $\lambda_{17} = \lambda_{18} = C_3C_4$

$\Phi_1 = \Omega_1 t + \nu_1$, $\Phi_2 = \Omega_2 t + \nu_2$

$\Phi_3 = 3\Omega_1 t + \nu_3$, $\Phi_4 = 3\Omega_2 t + \nu_4$

Like in the classical Duffing's equation with one exciting force, the unstable regions which occur due to the terms: $\cos 2\Omega_1$, $\sin 2\Omega_1$, $\cos 2\Omega_2$ and $\sin 2\Omega_2$ are called the first order unstable regions. There are two kinds of these regions, one in the neighbourhood of $\Omega_1 \approx \sqrt{\lambda_0}$, and the second in the neighbourhood of $\Omega_2 \approx \sqrt{\lambda_0}$.

In the same way we can define the third order unstable regions of two kinds for $\Omega_1 \approx \sqrt{\lambda_0}/3$ and $\Omega_2 \approx \sqrt{\lambda_0}/3$.

In this example, the particular attention will be given to the combined unstable regions, which occur due to the terms: $\cos(\Phi_1 \pm \Phi)_2$, $\cos(\Phi_1 \pm \Phi)_4$, $\cos(\Phi_2 \pm \Phi)_3$ and $\cos(\Phi_3 \pm \Phi)_4$. There are eight such regions and they take place in the neighbourhoods of:

- the first kind:

$$\overline{\Omega}_1 = (\Omega_1 + \Omega_2)/2 \approx \sqrt{\lambda_0}$$

- the second kind:

$$\bar{\Omega}_2 = (\Omega_1 - \Omega_2)/2 \approx \sqrt{\lambda_0}$$

- the third kind:

$$\bar{\Omega}_3 = (\Omega_1 + 3\Omega_2)/2 \approx \sqrt{\lambda_0}$$

- the fourth kind:

$$\bar{\Omega}_4 = (\Omega_1 - 3\Omega_2)/2 \approx \sqrt{\lambda_0}$$

- the fifth kind:

$$\bar{\Omega}_5 = (3\Omega_1 + \Omega_2)/2 \approx \sqrt{\lambda_0}$$

- the sixth kind:

$$\bar{\Omega}_6 = (3\Omega_1 - \Omega_2)/2 \approx \sqrt{\lambda_0}$$

- the seveth kind:

$$\bar{\Omega}_7 = 3(\Omega_1 + \Omega_2)/2 \approx \sqrt{\lambda_0}$$

- the eighth kind:

$$\bar{\Omega}_8 = 3(\Omega_1 - \Omega_2)/2 \approx \sqrt{\lambda_0}$$

The first approximate solution in the combined unstable region of the above kinds according to Floquet theory is :

- for the first kind region

$$\delta(t) = e^{\varepsilon_1 t}\, b^{(1)}\left(\cos \frac{\Omega_1 + \Omega_2}{2} t + \nu^{(1)}\right) \qquad (10.10a)$$

- for the second kind region

$$\delta(t) = e^{\varepsilon_2 t}\, b^{(2)} \cos\left(\frac{\Omega_1 + \Omega_2}{2} t + \nu^{(2)}\right) \qquad (10.10b)$$

- for the third kind region

$$\delta(t) = e^{\varepsilon_3 t}\, b^{(3)} \cos\left(\frac{\Omega_1 + 3\Omega_2}{2} t + \nu^{(3)}\right) \qquad (10.10c)$$

- for the fourth kind region

$$\delta(t) = e^{\varepsilon_4 t}\, b^{(4)} \cos\left(\frac{\Omega_1 - 3\Omega_2}{2} t + \nu^{(4)}\right) \qquad (10.10d)$$

- for the fifth kind region

$$\delta(t) = e^{\varepsilon_5 t}\, b^{(5)} \cos\left(\frac{3\Omega_1 + \Omega_2}{2} t + \nu^{(5)}\right) \qquad (10.10e)$$

- for the sixth kind region

$$\delta(t) = e^{\varepsilon_6 t}\, b^{(6)} \cos\left(\frac{3\Omega_1 - \Omega_2}{2} t + \nu^{(6)}\right) \qquad (10.10f)$$

- for the seventh kind region

$$\delta(t) = e^{\varepsilon_7 t}\, b^{(7)} \cos\left(\frac{3(\Omega_1 + \Omega_2)}{2} t + \nu^{(7)}\right) \qquad (10.10g)$$

- for the eighth kind region

$$\delta(t) = e^{\varepsilon_8 t}\, b^{(8)} \cos\left(\frac{3(\Omega_1 - \Omega_2)}{2} t + \nu^{(8)}\right) \qquad (10.10h)$$

where $\varepsilon_i > 0$, $i=1,2,\dots,8$. At the stability limit we have $\varepsilon_i=0$. To determine the boundaries of the unstable regions we insert the solutions (10.10) into variational equation (10.8) and conditions of nonzero solution for $b^{(i)}_{1/2}$ give us the following criteria to be satisfied at the stability limits of the

- first (+) and the second (–) kind:

$$[-(\frac{\Omega_1 \pm \Omega_2}{2})^2 + \lambda_0]^2 + \frac{(\Omega_1 \pm \Omega_2)^2}{4} a^2 - \frac{1}{4}\lambda_{11}^2 = 0 \qquad (10.11a)$$

- third (+) and the fourth (−) kind

$$[-(\frac{\Omega_1 \pm 3\Omega_2}{2})^2 + \lambda_0]^2 + \frac{(\Omega_1 \pm 3\Omega_2)^2}{4} a^2 - \frac{1}{4}\lambda_{13}^2 = 0 \qquad (10.11b)$$

- fifth (+) and the sixth (−) kind

$$[-(\frac{3\Omega_1 \pm \Omega_2}{2})^2 + \lambda_0]^2 + \frac{(3\Omega_1 \pm \Omega_2)^2}{4} a^2 - \frac{1}{4}\lambda_{15}^2 = 0 \qquad (10.11c)$$

- seventh (+) and the eighth (−) kind

$$[-\frac{9(\Omega_1 \pm \Omega_2)^2}{2} + \lambda_0]^2 + \frac{9(\Omega_1 \pm \Omega_2)^2}{4} a^2 - \frac{1}{4}\lambda_{17}^2 = 0 \qquad (10.11d)$$

To investigate the existence of combined period doubling bifurcation (Iooss and Joseph (1980)) at the critical points $\overline{\Omega}^i_{1,2}$ which are the solutions of equations (10.11) the existence and stability of the steady-state solution of the complete variational equation should be examined.

As all equations (10.11) have the same form:

$$[(\overline{\Omega}^{(i)})^2 + \lambda_0]^2 + (\overline{\Omega}^{(i)})^2 a^2 - \frac{1}{4} \lambda_{1r} = 0 \qquad (10.11)$$

where r = { 1 for the first and the second kind
3 for the third and the fourth kind
5 for the fifth and the sixth kind
7 for the seventh and the eight kind

similar to the equation obtained in Szemplińska-Stupnicka and Bajkowski (1986). Following their calculations we obtain the following formulas for the amplitude of the bifurcating solutions:

$$b^{(i)} = \frac{1}{3}\left[1 - \frac{a^2}{2(\lambda_0 - (\bar{\Omega}_1^{(i)})^2)}\right]\left[\bar{\Omega}^{(i)} - \bar{\Omega}_1^{(i)}\right]$$

or (10.12)

$$b^{(i)} = \frac{1}{3}\left[1 - \frac{a^2}{2((\bar{\Omega}_2^{(i)})^2 - \lambda_0)}\right]\left[\bar{\Omega}_2^{(i)} - \bar{\Omega}^{(i)}\right]$$

Making use of the Routh-Hurwitz criterion the following conditions for the stability of the solutions (10.10) for $\varepsilon_i = 0$ have been obtained:

$$\frac{3}{2}\frac{b^{(i)}}{\bar{\Omega}^i}\left[(\Omega_{1,2}^{(i)})^2 - \lambda_0\right] > 0 \qquad (10.13)$$

From equations (10.12) we have that:

$$(\Omega_1^{(i)})^2 - \lambda_0 < 0 \quad \text{and} \quad (\Omega_2^{(i)})^2 - \lambda_0 > 0 \qquad (10.14)$$

the above conditions show that all these bifurcation solutions (10.10) are stable in the neighbourhood of $\bar{\Omega}_1^i$ and unstable in the neighbourhood of $\bar{\Omega}_2^i$.

In the same method the existence of the further combined bifurcations can be shown and finally one obtains the stable bifurcation diagram shown in Figure 10.3.

The formulas (10.11) allow to calculate the value of amplitude and frequency at the stability limit of the above discribed eight kinds. The example of these calculations for bifurcations of first four kinds is shown in

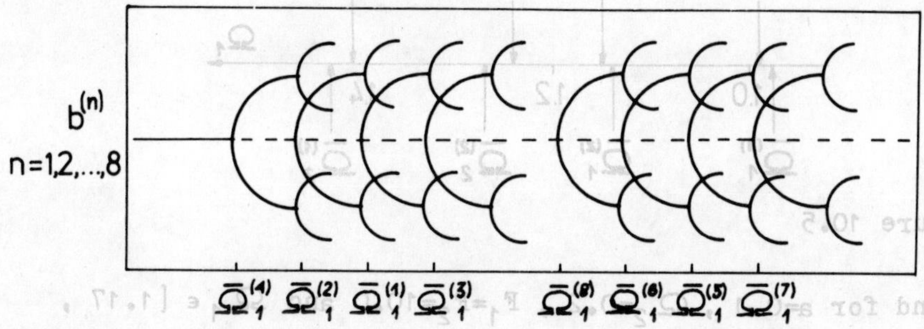

Figure 10.3

in Figure 10.4 .

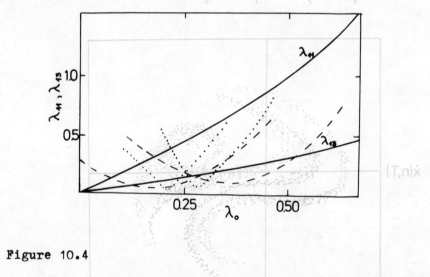

Figure 10.4

The bifurcation $(\Omega_1 + \Omega_2)/2$ takes place between $\bar{\Omega}_1^{(1)}$ and $\bar{\Omega}_2^{(1)}$, bifurcation $(\Omega_1 - \Omega_2)/2$ between $\bar{\Omega}_1^{(2)}$ and $\bar{\Omega}_2^{(2)}$. Bifurcations $(\Omega_1 \pm 3\Omega_2)/2$ take place between $\bar{\Omega}_1^{(3)}$ and $\bar{\Omega}_2^{(3)}$ and between $\bar{\Omega}_1^{(4)}$ and $\bar{\Omega}_2^{(4)}$. In the zone between $\bar{\Omega}_1^{(4)}$ and $\bar{\Omega}_2^{(2)}$ all bifurcations of first four kinds take place - - Figure 10.5 . In this zone the chaotic behaviour has been

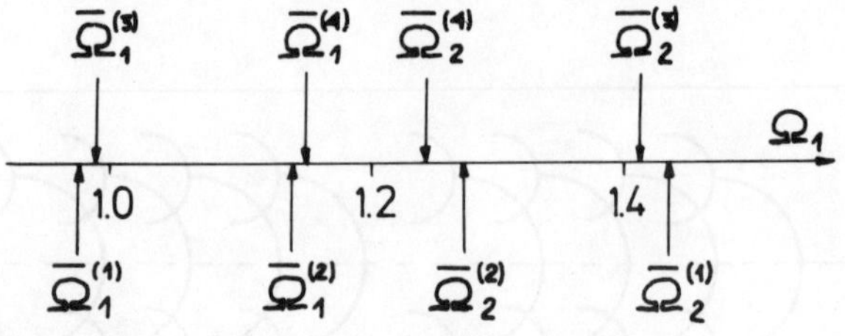

Figure 10.5

found for a=0.1 , Ω_2=0.2 , F_1=F_2=10.0 and $\Omega_1 \in [1.17 , 1.20]$. In Figure 10.6 (a)-(b) Poincare maps of this system have been shown. In Figure 10.7 the frequency spectra of this system have been presented. For Ω_1=1.22 this spectrum consists of amplitudes with the following frequencies: Ω_1 , Ω_2 , $3\Omega_1$, $3\Omega_2$ and combined subharmonic

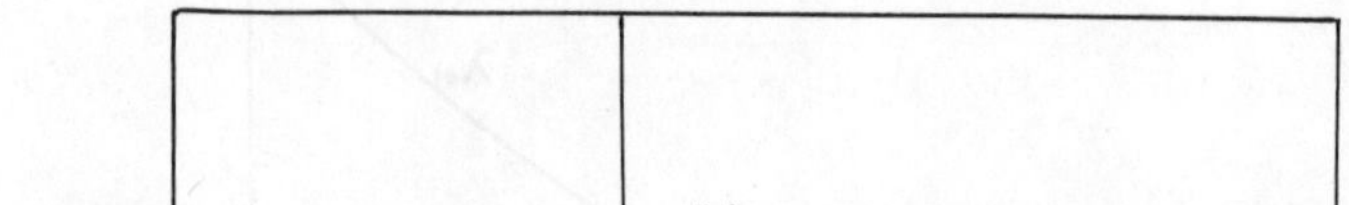

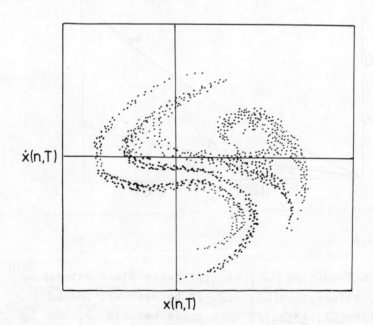

(a) Ω_1 = 1.17

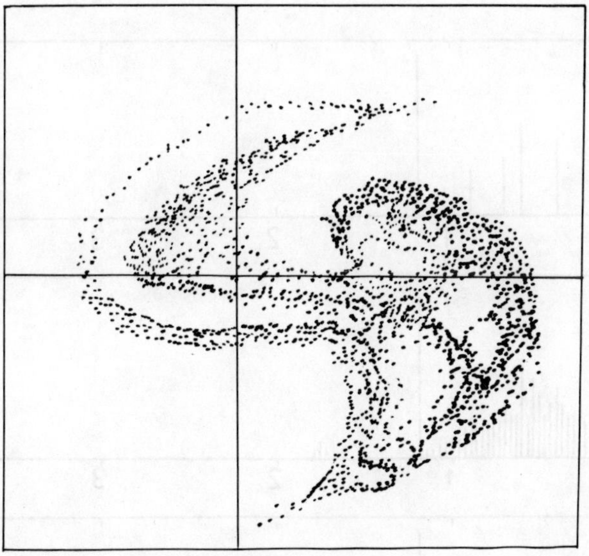

(b) $\Omega_1 = 1.20$
Figure 10.6

components with frequencies:

$(\Omega_1 + \Omega_2)/2$

$(\Omega_1 - \Omega_2)/2$

$(\Omega_1 + 3\Omega_2)/2$

$(\Omega_1 - 3\Omega_2)/2$

- Figure 10.7(a). Next for Ω_1=1.20 the spectrum becomes continous in the neighbourhoods of combined subharmonic components - Figure 10.7(b). The chaotic behaviour is still visible for Ω_1=1.17 - Figure 10.7(c). For the

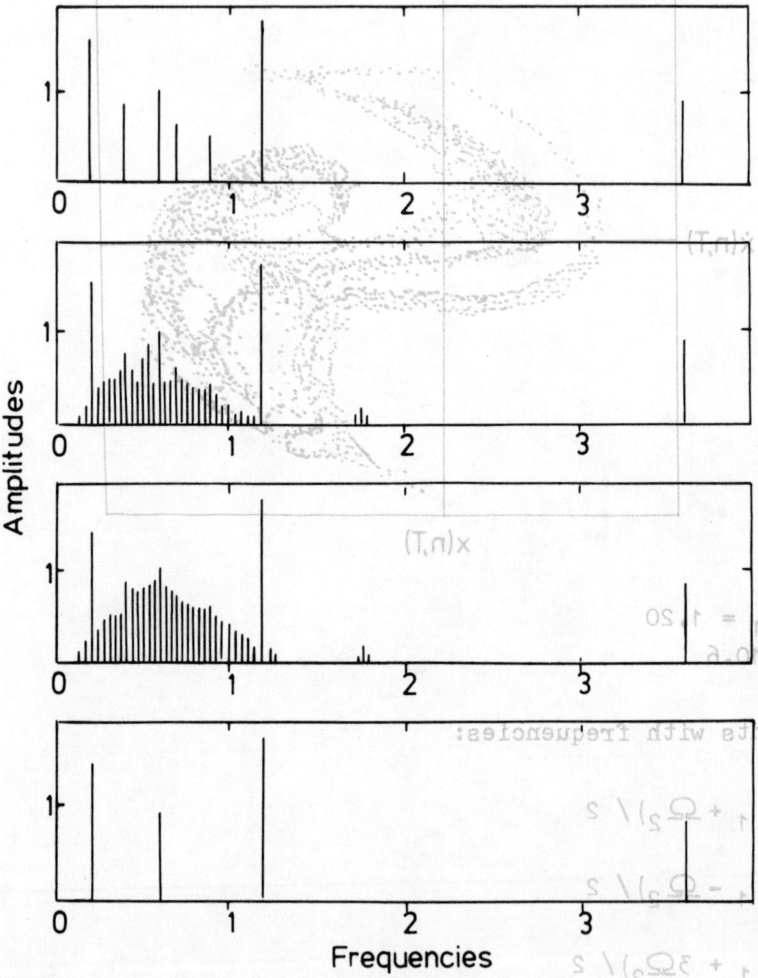

Figure 10.7

decreasing Ω_1 we observe the spectrum consisting of the components with frequencies : Ω_1 , Ω_2 , $3\Omega_1$ and $3\Omega_2$ for Ω_1= 1.15 - Figure 10.7 (d).

10.2.2. Stochastic case

Now let's consider the influence of the noise on the transition to chaos in the system (10.1).

The similar power spectra as in Figure 10.7 for

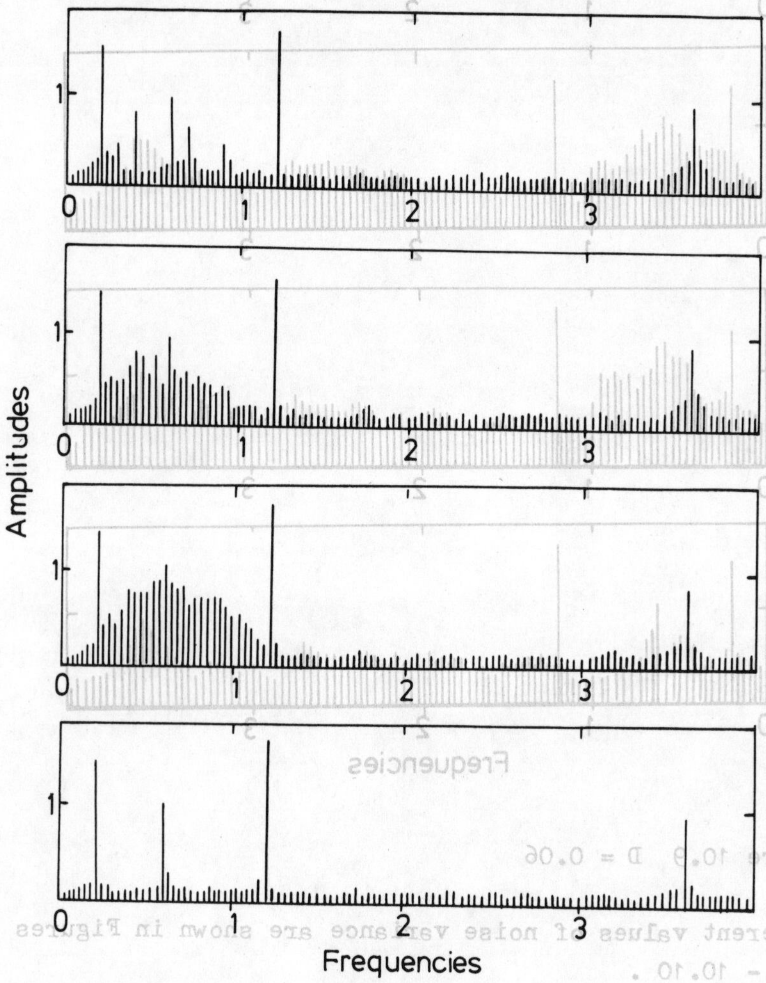

Figure 10.8 D = 0.02

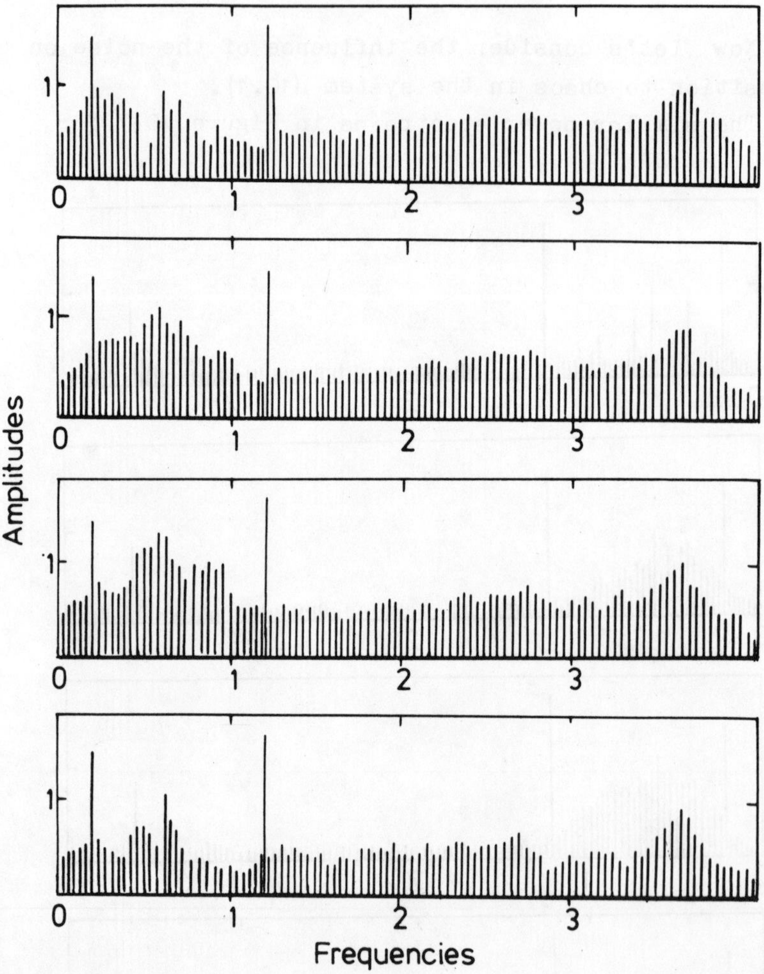

Figure 10.9 D = 0.06

different values of noise variance are shown in Figures 10.8 - 10.10 .

As the continuous frequency spectrum is characteristic for all stochastic systems we cannot state from this

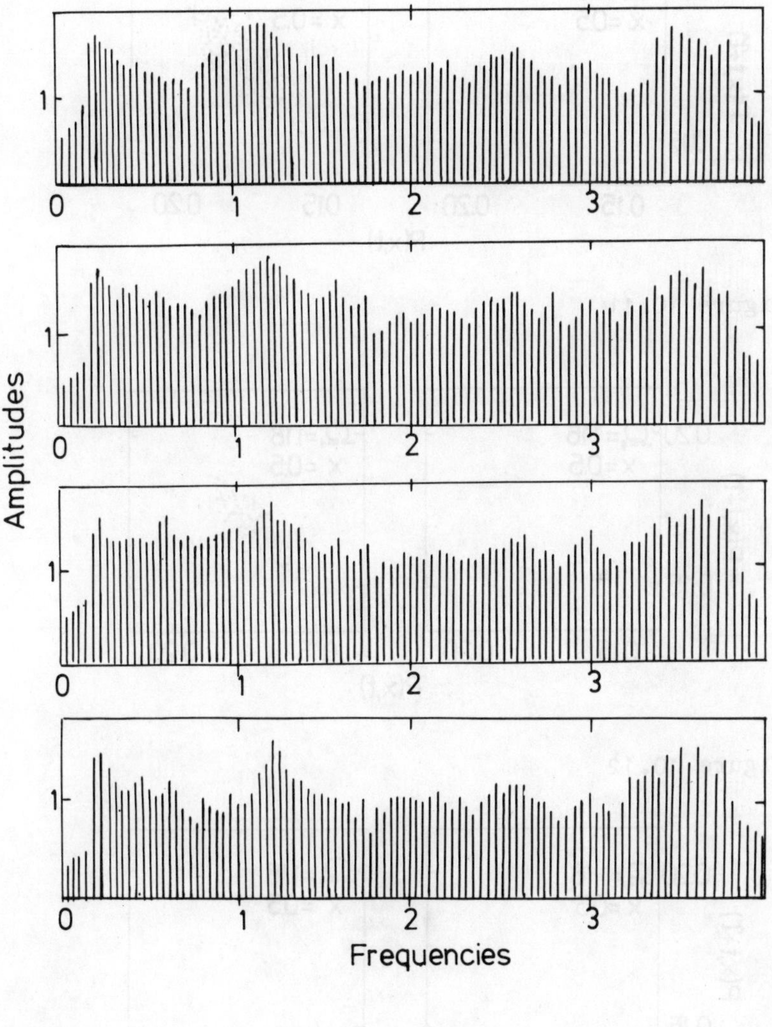

Figure 10.10 D = 0.1

if the response of the system is a chaotic or a stochastic process. In Figures 10.11 - 10.13 maps P(x, t) versus P(x, t+τ) for constant x and τ are shown. As it was shown

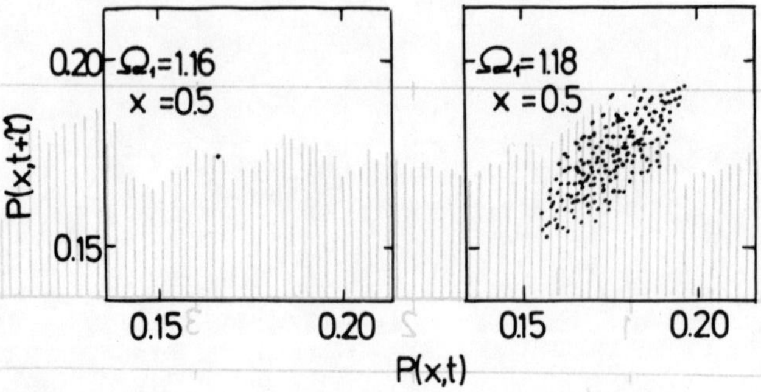

Figure 10.11

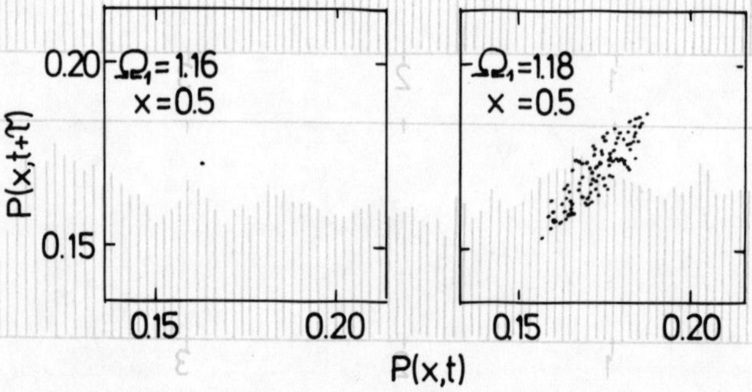

Figure 10.12

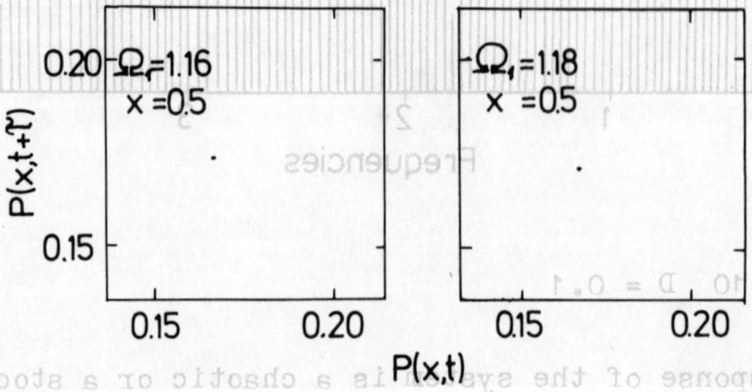

Figure 10.13

in Chapter 5 these maps allow to distinquish between regular and chaotic stochastic processes. In our example, we find that the zone, where chaotic behaviour takes place $\Omega_1 \in [1.17, 1.20]$ decreases if we have noise in the system. The dependence of this zone on the noise variance is shown in Figure 10.14 .

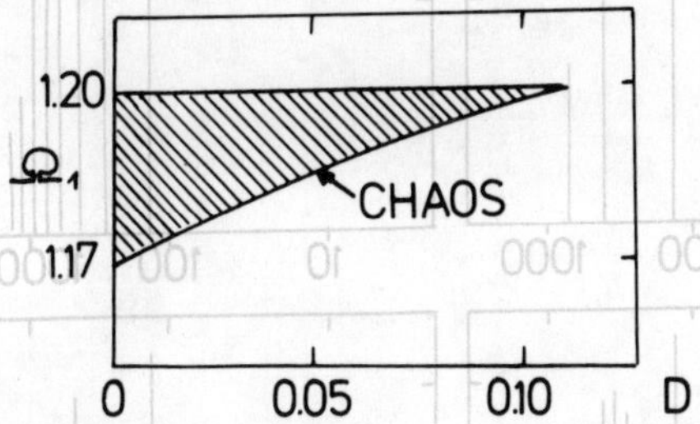

Figure 10.14

To summarize we can say that the route to chaos in the noisy system has the same form as in the deterministic system. It is especially visible for the noise with small variances where we can distinquish the same combined bifurcation as in the deterministic case. For the bigger values of noise variance the chaotic zone decreases and finally disappears.

10.3. Feedback control system with time delay

As the last example we consider the route from regular motion regular stochastic process to the chaotic

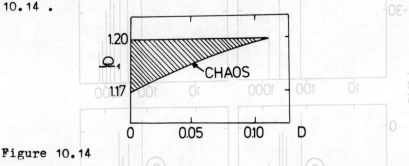

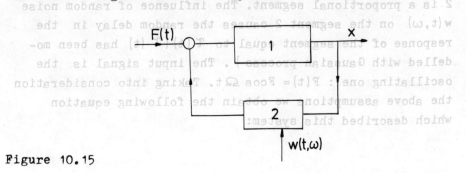

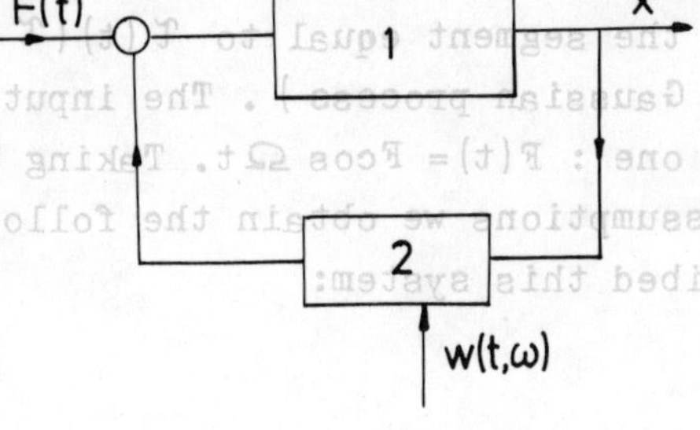

Figure 10.15

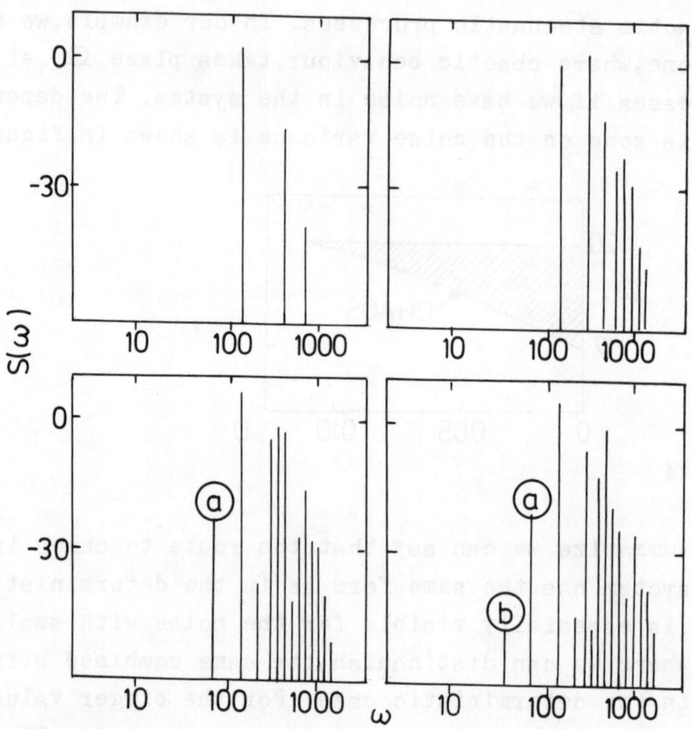

Figure 10.16

motion(chaotic stochastic process) in the automatic control feedback system shown in Figure 10.15 .

The segment 1 is a non-linear oscillating segment and 2 is a proportional segment. The influence of random noise $w(t,\omega)$ on the segment 2 causes the random delay in the response of the segment equal to $\tau(t)$ ($\tau(t)$ has been modelled with Gaussian process) . The input signal is the oscillating one : $F(t) = F\cos \Omega t$. Taking into consideration the above assumptions we obtain the following equation which described this system:

$$\ddot{x}(t) + \alpha \dot{x}(t) + kx(t - \tau(t)) + \eta x^3 t = F\cos\Omega t$$

Transition from regular response to chaotic one in the system without time delay has been investigated by Novak and Frehlich (1982) for $\alpha/k = 0.1$, $\Omega = 144$, $\eta/k^2 = 0.5$, $k = 109$ and different values of F . The obtained transition to chaos is described in Figure 10.16 showing the power spectra of the system. With the increase of the parameter F in the input to the system we obtain bifurcation solution

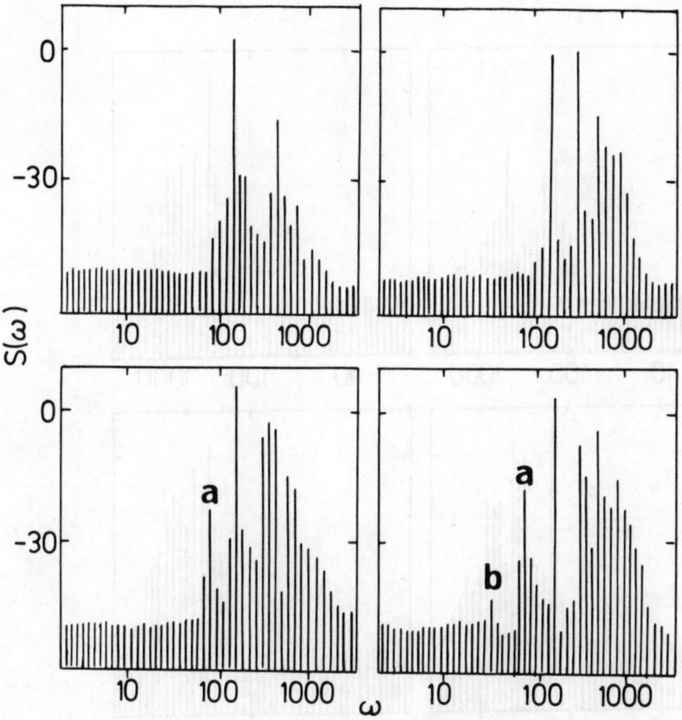

Figure 10.17 D = 0.1

with double period ⓐ and fourfold period ⓑ of the input signal frequency.

In this example we describe the influence of random time delay $\tau(t)$ on the route from regular stochastic process to the chaotic stochastic process. In Figure 10.17 and 10.18 the same power spectra of the noisy system have been shown (the variance of the time delay D=0.1 - Figure 10.17 and D=0.2 - Figure 10.18).

The comparison of Figures 10.17 , 10.18 and 10.16 shows that the mechanism of the transition to chaos in

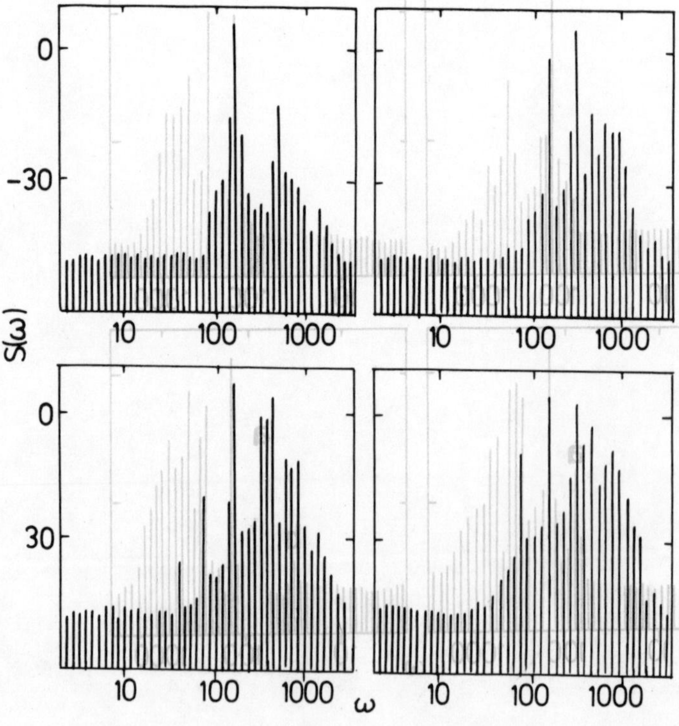

Figure 10.18 D = 0.2

noisy system is the same as in deterministic one. In noisy power spectra the maxima ⓐ and ⓑ which indicate the existence of the same bifurcations as in deterministic case are still visible. Moreover the response of the noisy system becomes chaotic for the value $F = 32.4$, the same value as in deterministic system.

The analisis of the influence of deterministic time delay on the chaotic behaviour of Duffing's oscillators has been presented in Kapitaniak (1987).

APPENDIX

While considering many practical problems , stochastic processes are described as sums of harmonics with given deterministic frequencies and random amplitudes and phases, for example:

$$\eta_R(t,\omega) = \sum_{k=1}^{N} A_k \cos(\nu_k t + \varphi_k) \tag{A1}$$

where ν_k are constant , A_k and φ_k are random variables. In this description stochastic process $\eta(t,\omega)$ is a sum of N harmonic oscillations with amplitudes A_k and frequencies ν_k . The set of all frequencies ν_k ($k=1,2,..,N$) as a set of points in the interval $(-\infty, \infty)$ is called a spectral density of $\eta(t,\omega)$.

In computer simulations of random process A1 we consider only nonnegative spectral density and in many methods we consider only finite sets of frequencies $[\nu_{min}, \nu_{max}]$.

In that way we can approximate for example band-limited white noise i.e.: the process with spectral density:

$$S_0(\nu) = \begin{cases} \dfrac{D^2}{\nu_{max} - \nu_{min}} & \text{for } \nu \in [\nu_{min}, \nu_{max}] \\ 0 & \text{for } \nu \notin [\nu_{min}, \nu_{max}] \end{cases} \tag{A2}$$

where D is a variance of $\eta(t,\omega)$.

There are a few methods of calculating the coefficients of formula(A1) . In all of them φ_k are independent random variables with uniform distribution on the interval $[0, 2\pi]$ and A_k and ν_k can be calculated from the following formulas:

1. Borgman's method (Borgman (1969))

In this method, the amplitudes of all harmonics are constant and equal to:

$$A_k = \sqrt{\frac{2D^2}{N}} \tag{A3}$$

ν_k are independent random variables with uniform distribution:

$$p_k(\nu_k) = p(\nu) = \frac{S_o(\nu)}{D^2} \tag{A4}$$

2. Rice method (Shinczuka (1977))

Amplitudes A_k and frequencies ν_k are deterministic and given by:

$$A_k = \sqrt{2S_o(\nu_k)\Delta\nu} \tag{A5}$$

$$\nu_k = (k - 0.5)\Delta\nu + \nu_{min} \tag{A6}$$

$$\Delta\nu = \frac{\nu_{max} - \nu_{min}}{N} \tag{A7}$$

3. Shinczuka's method (Shinczuka (1977))

In this method only amplitudes A_k are deterministic and given by (A5). The frequencies ν_k are random and described by the formula:

$$\nu_k = (k - 0.5)\Delta\nu + \delta\nu_k + \nu_{min} \tag{A8}$$

where $\Delta\nu$ is given by (A7) and $\delta\nu_k$ are independent random variables uniformly distributed in the interval $[-\Delta\nu/2, \Delta\nu/2]$.

Single realization of the process $\eta(t,\omega)$ is obtained by drawing the random variables φ_k and $\delta\nu_k$ and substituting them into the equation (A1).

As the measure of the quality of these realizations of $\eta_R(t,\omega)$ the mean square error ε of the spectral density of the process $\eta(t,\omega)$ - $S_0(\nu)$ and its approximation $\eta_R(t,\omega)$ spectral density - $S(\nu)$

$$\varepsilon = \frac{1}{\nu_{max} - \nu_{min}} \left[\int_{min}^{max} [S(\nu) - S_0(\nu)]^2 \, d\nu \right]^{\frac{1}{2}} \qquad (A9)$$

can be taken (Wróbel (1985)).

To obtain a good approximation of the band-limited white noise the number of harmonics must be relevantly big ($N > 30$). For all simulations presented in this book ε was less than 0.1 .

REFERENCES

Argoul, F. and Arneodo, A., J. Mecanique Theorique et Appliquee , Numero Special, 241, (1984)

Argoul, F., Arneodo, A., Collet, P. and Lesne, A., Europhys. Lett., 3, 643, (1987)

Arneodo, A., Coullet, P., and Tresser, C., Phys. Lett. 70A, 74, (1979)

Arneodo, A., Coullet, P. and Tresser, C., Phys. Lett. 79A, 59, (1980)

Arneodo, A., Coullet, P. and Tresser, C., Phys. Lett. 81A, 197, (1981)

Arneodo, A., Coullet, P. and Tresser, C., J. Stat. Phys. 27, 171, (1982)

Arneodo, A., Coullet, P. and Spiegel, E. A., Phys. Lett. 92A, 369, (1982)

Arnold, V.I., Mathematical Methods of Classical Mechanics, Springer (1980)

Atmanspacher, H. and Scheingraber, H., Found. Phys. in press, (1987)

Barrow, J.D., Phys. Reports 85, 1, (1982)

Becker, K.-H. and Seydey, R., Lecture Notes in Mathematics 878, 98 (1982)

Benettin, G., Cercignani, C., Galgani, L., and Giorgilli, A. Phys. Rev. A19, 431, (1979)

Borgman, L. E., J. Waterways and Harbours Division, 95, 557, (1969)

Brahic, A., Astron. and Astrophys. 12, 98 (1971)

Cesari, L., Asynptotic Behaviour and Stability Problems in Ordinary Differential Equations , Springer (1973)

Chang, S. J., and Wright, J., Phys. Rev. A23, 1419 (1981)

Chirikov, B. V., and Shepelyansky, D. L., Pis'ma JETP 34, 171, (1981)

Chirikov, B. V., Phys. Peports 52, 263, (1979)

Chrostowski, J., Vallee, R., and Delisle, C., Can. J. Phys.

61, 1143 (1983
Chui, S. T., and Ma, K. B., Phys. Rev. A26, 2262 (1982)
Coullet, P., and Tresser, C., J. de Phys. Lett. 41, L255 (1980)
Coullet, P., and Tresser, C., J. Mecanique Theorique et Appliquee, Numero Special, 217 (1984)
Coullet, P., and Tresser, C., J. de Phys. 39, Coll. C5 (1978)
Coullet, P., Tresser, C., and Arneodo, A., Phys. Lett. 72A, 268 (1979)
Coullet, P., and Vanneste, C., Helv. Phys. Acta 56, 813 (1983)
Crutchfield, J.P., and Huberman, B. A., Phys. Lett. 77A, 407 (1980)
Crutchfield, J.P., Farmer, J. D., Packard, N. H., and Shaw, R. S., Phys. Lett. 76A, 1, (1980)
Crutchfield, J.P., Nauenberg, M., and Rudnick, J., Phys. Rev. Lett. 46, 933 (1981)
Crutchfield, J.P., Farmer, J. D., and Huberman, B. A., Phys. Reports 92,45 (1982)
Crutchfield, J.P., and Packard, N. H., Physica 7D, 201 (1983)
Crutchfield, J.P., and Packard, N. H., Int. J. Theor. Phys. 21, (1982)
Curry, J. H., Commun. Math. Phys. 60, 193 (1978)
Curry, J. H., J. Stat. Phys. 26, 683 (1981)
Cvitanovic, P., and Myrheim, J., Phys. Lett. 94A, 329 (1983)
Cvitanovic, P., Universality in Chaos, Adam Hilger Ltd, Bristol (1986)
Daido, H., Prog. Theor. Phys. 69, 1304 (1983)
Daido, H., Prog. Theor. Phys. 70, 879 (1983)
D'Humieres, D., Beasley, M. R., Huberman, B. A., and Libchaber, A., Phys. Rev. A26, 3483 (1982)
Doob, J. L., Stochastic Processes, John Wiley, New York

(1953)
Dowell, E. H., J. Sound Vibr. 85, 333 (1982)
Dowell, E. H., J. Appl. Mech. 51, 664 (1984)
Dowell, E. H., and Pereshki, C., J. Appl. Mech. 53, 5 (1986)
Ebeling, W., Herzel, H., Richert, W., and Schimansky-Geier, ZAMM 66, 141 (1986)
Eckmann, J.-P., Rev. Mod. Phys. 53, 643 (1981)
Eckmann, J.-P., Thomas, L. E., J. Phys. A15, L261 (1982)
Eckmann, J.-P., Thomas, L. E., and Wittwer, P., J. Phys. A14, 3153 (1981)
Farmer, J. D., Crutchfield, J.P., Froehling, H., Packard, N. H., and Shaw, R. S., Annals N.Y. Acad. Sci. 357, 453 (1980)
Farmer, J. D., Phys. Rev. Lett. 47, 179 (1981)
Farmer, J. D., Physica 4D, 366 (1982)
Farmer, J. D., Z. Naturforsch. 37a, 1304 (1982)
Farmer, J. D., Ott, E., and Yorke, J. A., Physica 7D, 153 (1983)
Farmer, J. D., Hart, J., and Weidman, P., Phys. Lett. 91A, (1982)
Feigebaum, M. J., J. Stat. Phys. 19, 25 (1978)
Feigebaum, M. J., J. Stat. Phys. 21, 669 (1979)
Feigebaum, M. J., Los Alamos Sci. Summer , 4 (1980)
Feigebaum, M. J., Phys. Lett. 74A, 375 (1980)
Feigebaum, M. J., and Hasslacher, B., Phys. Rev. Lett. 49, (1982)
Feigebaum, M. J., Physica 7D, 16 (1983)
Feingold, M., and Perez, A., Physica 9D, 433 (1983)
Ford, J., The Statistical Mechanics of Classical Analytic Dynamics, in Fundamental Problems in Statistical Mechanics, North-Holland, Amsterdam (1985)
Ford, J., and Lunsford, G.H., Phys. Rev. A1, 59 (1970)
Fraser, S., and Kapral, R., Phys. Rev. A23, 3303 (1981)
Fraser, S., and Kapral, R., Phys. Rev. A25, 2827 (1982)

Froehling, H., Crutchfield, J. P., Farmer, J. D., and Shaw, R. S., Physica 3D, 605 (1981)
Froeschle, C., J. Mecanique Theorique et Applique, Numero special, 101 (1984)
Froyland, J., Phys. Lett. 97A, 8 (1983)
Fujisaka, H., Prog. Theor. Phys. 68, 1105 (1982)
Fujisaka, H., and Grossman, S., Z. Phys. B48, 261 (1982)
Fujisaka, H., Prog. Theor. Phys. 70, 1264 (1983)
Gollub, J. P., and Swinney, H. L., Phys. Rev. Lett. 35, 927 (1975)
Gollub, J. P., Brunner, T. O., and Danly, B. G., Science 200, 48 (1978)
Grassberger, P., and Procaccia, I., Physica 9D, 189 (1983)
Grassberger, P., and Procaccia, I., Phys. Rev. Lett. 50, (1983)
Grassberger, P., and Procaccia, I., Phys. Rev. A28, 2591 (1983)
Grossmann, S., Phys. Lett. 97A, 263 (1983)
Guckenheimer, J., Nature 298, 358 (1982)
Guckenheimer, J., Nonlinear Oscillations, Dynamical Systems and Bifurcations of Vector Fields, Springer (1983)
Haken, H., ed., Chaos and Order in Nature, Springer (1981)
Hao, B.-L., ed., Chaos , World Scientific, Singapore (1984)
Henon, M., and Heiles, C., Astron. J., 69, 73 (1964)
Herzel, H.-P., and Pompe, B., Phys. Lett. 122A, 121 (1987)
Herzel, H.-P., Ebeling, W., and Schulmeister, Th., Z. Naturforsch. 42, 136 (1987)
Herzel, H.-P., Ebeling, W., Schimansky-Geier, L., and Sel'kov, E. E., The Influence of Noise on an Biochemical Oscillator, in Ebeling, W., and Peschel, M., Lotka-Volterra-Approach to Cooperation and Competition in Dynamic Systems, Akademie-Verlag, Berlin (1985)
Holmes, P., and Marsden, J. E., Arch. Rat. Mech. Anal. 76, 135 (1981)

Holmes, P., Phyl. Trans. Roy. Soc. A293, 420 (1979)

Holmes, P., and Holmes, C., J. Sound Vibr. 78,161 (1981)

Holmes, P., and Rand, D. A., Quart. Appl. Math. 35, 495 (1978)

Holmes, P., and Whitley, D., Physica 7D, 111 (1983)

Hu, B., Phys Reports 91, 233 (1982)

Huberman, B. A., Crutchfield, J. P., and Packard, N. H., Appl. Phys. Lett. 45, 750 (1980)

Huberman, B. A., and Rudnick, J., Phys. Rev. Lett. 45, 154 (1980)

Huberman, B. A., Physica 18A, 323 (1983)

Huberman, B. A., and Zisook, A. B., Phys. Rev. Lett. 46, 626 (1981)

Hunt, E. R., Phys. Rev. Lett. 49, 1054 (1982)

Iooss, G., and Langford, W. F., Ann. N. Y. Acad. Sci. 357, 489 (1980)

Iooss, G., and Joseph, D. D., Elementary Stability and Bifurcation Theory, Springer (1980)

Iooss, G., Helleman, R. H. G., and Stora, R., Chaotic Behaviour of Deterministic Systems, Les Houches, North-Holland (1983)

Ito, A., Prog. Theor. Phys. 61, 45 (1979)

Ito, A., Prog. Theor. Phys. 62, 620 (1979)

Kadanoff, L. P., Phys. Today December, 46 (1983)

Kai, T., and Tomita, K., Prog. Theor. Phys. 64, 1532 (1980)

Kai, T., J. Stat. Phys. 29, 329 (1982)

Kaneko, K., Prog. Theor. Phys. 69, 1427 (1983)

Kaplan, H., Phys. Lett. 97A, 365 (1983)

Kapitaniak, T., J. Sound Vibr. 102, (1985)

Kapitaniak, T., J. Sound Vibr. 107, 177 (1986)

Kapitaniak, T., Phys. Lett. 116A, 251 (1986)

Kapitaniak, T., J. Sound Vibr. 114, 588 (1987)

Kapitaniak, T., Awrejcewicz, J., and Steeb, W.-H., J. Phys. A20, L355 (1987)

Kapitaniak, T., J. Soc. Jpn 56, 1951 (1987)

Kapral, R., Schell, M., and Fraser, S., J. of Phys. Chem. 86, 2205 (1982)
Karney, C. F. F., Rechester, A. B., and White, R. B., Physica 4D, 425 (1982)
Karney, C. F. F., Physica 8D, 360 (1983)
Katok, A. B., Publ. Math. IHES. 51, 137 (1980)
Knobloch, E., Phys. Lett. 82A, 439 (1981)
Kozak, J. J., Musho, M. K., and Hatlee, M. D., Phys. Rev. Lett. 49, 1801 (1982)
Lafon, A., Rossi, A., and Vidal, C., J.Physique 44, 505 (1983)
Landa, P. S., Stratonovich, R. L., Sov. Phys.-Dokl. 27, 1032 (1982)
Landau, L. D., C. R. Acad, Sci. USSR, 44, 311 (1944)
Lanford, O. E., Annual Review of Fluid Mechanics 14, 347 (1982)
Lanford, O. E., Physica 7D, 124 (1983)
Lapunov, A. M., Ann. Math. Study 17 (1947)
Lauterborn, W., Acoustic Turbulence , in Frontier in Physical Acoustics, Soc. Italiana di Fisica, Bologna (1986)
Lauterborn, W., and Cramer, E., Phys. Rev. Lett. 47, 1445 (1981)
Lauterborn, W., Appl. Sci. Res. 38, 165 (1982)
Leven, R. W., Pompe, B., Wilke, C., and Koch, B. P., Physica 16D, 371 (1985)
Leven, R. W., and Koch, B. P., Phys. Lett. 86A, 71 (1981)
Levi, M., Periodically forced relaxation oscillations, in Lect. Notes in Math. 819, 300 (1980)
Levi, M., Memoirs Am. Math. Soc. 244, (1981)
Levy, Y. E., Phys. Lett. 88A, 1 (1982)
Libchaber, A., Laroche, C., and Fauve, S., Physica 7D, 73 (1983)
Libchaber, A., Laroche, C., and Fauve, S., J. de Phys. Lett. 43, L211 (1982)

Linsay, P. S., Phys. Rev. Lett. 47, 1349 (1981)
Lorenz, E. N., J. Atmos. Sci. 20, 130 (1963)
Lorenz, E. N., Tellus 16, 1 (1964)
Lorenz, E. N., Ann. N. Y. Acad. Sci. 357, 282 (1980)
Lorenz, E. N., J. Atmos. Sci. 36, 1685 (1980)
Lorenz, E. N., Physica 17D, 279 (1985)
Lorenz, E. N., Physica 8D, 90 (1984)
Lunsford, G. H., and Ford, J., J. Math. Phys. 13, 700 (1972)
Malomed, A. B., Physica 8D, 343 (1983)
Mandel, P., and Kapral, R., Opt. Commun. 47, 151 (1983)
Manneville, P., and Pomeau, Y., Phys. Lett. 75A, 1 (1979)
Manneville, P., Phys. Lett. 79A, 33 (1980)
Manneville, P., J. de Phys. 41, 1235 (1980)
Manneville, P., and Pomeau, Y., Physica 1D, 219 (1980)
Manneville, P., Phys. Lett. 90A, 327 (1982)
Marsden, E. J., and McCracken, M., The Hopf Bifurcation and its Applications, Springer (1976)
Martin, P. C., J. de Phys. 37, Colloq. C1 (1976)
Marzec, C. J., and Spiegel, E. A., SIAM J. Appl. Math. 38, 387 (1980)
Matsumoto, K., and Tsuda, I., J. Stat. Phys. 31, 87 1983 , Addendum, ibid, 33, 757 (1983)
Mayer-Kress, G., and Haken, H., Phys. Lett. 82A, 151 (1981)
Mayer-Kress, G., and Haken, H., J. Stat. Phys. 26, 149 (1981)
McGuinness, M. J., Phys. Lett. 99A, 5 (1983)
McLaughlin, J. B., J. Stat. Phys. 19, 587 (1980)
McLaughlin, J. B., J. Stat. Phys. 24, 375 (1981)
Melnikov, V. K., Trans. Moscow Math. Soc. 12, 1 (1963)
Millonni, P.W., Ackerhalt, J. P., and Galbraith, H. W., Phys. Rev. A28, 887 (1983)
Miracky, R. F., Clarke, J., and Koch, R. H., Phys. Rev. Lett. 50, 857 (1983)
Moloney, J. V., Hopf, F. A., and Gibbs, H. M., Phys. Rev. A25, 3442 (1982)

Moon, F. C., and Holmes, P. J., J. Sound Vibr. 65, 285 , 69, 285 (1980) (1979)

Moon, F. C., J. Appl. Mech. 47, 638 (1980)

Moon, F. C., and Shaw, S. W., Int. J. Non-Linear Mech. 18, 465 (1983)

Mori, H., and Fujisaka, H., Prog. Theor. Phys. 63, 1931 (1980)

Mori, H., Prog. Theor. Phys. 63, 1044 (1980)

Mori, H., and Ose, T., Prog. Theor. Phys. 66, 4 (1981)

Nagashima, T., and Shimada, I., Prog. Theor. Phys. 58, 1318 (1977)

Nagashima, T., and Haken, H., Phys. Lett. 96A, 385 (1983)

Nakamura, K., Prog. Theor. Phys. 57, 6 (1977)

Nakamura, K., Prog. Theor. Phys. 59, 64 (1978)

Nakamura, K., Suppl. Prog. Theor. Phys. 64, 378 (1978)

Nakamura, K., Proc. Inst. Nat. Sci. Nihon Univ. 14, 9 (1979)

Nakatsuka, H., Asaka, S., Itoh, H., Ikeda, K., and Matzuoka, M., Phys. Rev. Lett. 50, 109 (1983)

Nauenberg, M., and Rudnick, J., Phys. Rev. B24, 493 (1981)

Nicolis, J. S., Mayer-Kress, G., and Haubs, H., Z. Naturforsch. 38a, 1157 (1983)

Novak, S., and Frehlich, R. G., Phys. Rev. A26, 3660 (1982)

Ogura, H., Ueda, Y., and Yoshida, Y., Prog. Theor. Phys. 66, 2280 (1981)

Oono, Y., and Takahashi, Y., Prog. Theor. Phys. 63 1804 (1980)

Oono, Y., Kohda, T., and Yamazaki, H., J. Phys. Soc. Jpn. 48, 738 (1980)

Oseledec, V. I., Trans. Moscow Math. Soc. 19, 197 (1968)

Ostlund, S., Rand,D., Sethna, J., and Siggia, E., Physica 8D, 303 (1983)

Ott, E., Rev. Mod. Phys. 53, 655 (1981)

Ott, E., and Hanson, J. D., Phys. Lett. 85A, 20 (1981)

Ott, E., Yorke, E. D., and Yorke J. A., Physica 16D, 62 (1985)

Parlitz, U., and Lauterborn, W., Phys. Lett. 107A, 351 (1985)

Parlitz, U., and Lauterborn, W., Z. Naturforsch. 41a, 605 (1986)

Parlitz, U., and Lauterborn, W., Phys. Rev. A36, (1987)

Pederson, N. F., Soerenson, O. H., Dueholm, B., and Mygind, J., J. Low Temp. Phys. 38, 1 (1980)

Pederson, N. F., and Davidson, A., Appl. Phys. Lett. 39, 830 (1981)

Perez, J., and Jeffries, C., Phys. Rev. B26, 3460 (1982)

Perez, J., and Glass, L., Phys. Lett. 90A, 441 (1982)

Pesin, Ya. B., Russ. Math. Surv. 32:5, 55 (1977)

Piszczek, K., and Nizioł, J., Random Vibration of Mechanical Systems, PWN, Warsaw (1986)

Pikovsky, A. S., J. Phys. A16, L109 (1983)

Poincare, H., Les Methodes Nouvelles de la Mecanoque Celeste, Gauthier- Villars, Paris (1892)

Pomeau, Y. M., and Manneville, P., Commun. Math. Phys. 74, 189 (1980)

Pomeau, Y. M., Roux, J. C., Rossi, A., Bachelart, A., and Vidal, C., J. Phys. Lett. 42, 271 (1981)

Poppe, D., and Korsch, J., Physica 24D, 367 (1987)

Rabinovich, M., Ann. N. Y. Acad. Sci. 375, 435 (1980)

Rabinovich, M., Sov. Phys. Usp. 21, 443 (1978)

Roberts, J. B., and Spanos, P.-T. D., Inter. J. Non-linear Mech. 21, 111 (1986)

Roberts, J. B., J. Sound Vibr. 50, 145 (1977)

Romeiras, F. J., and Ott, E., Phys. Rev. A35, 4404 (1987)

Rollins, R. W., and Hunt, E. R., Phys. Rev. Lett. 49, 1295 (1982)

Roux, J.-C., Simoyi, R. H., and Swinney, H. L., Physica 8D, 257 (1983)

Ruelle, D., and Takens, F., Commun. Math. Phys. 20, 167 (1971)

Ruelle, D., Publ. Math. IHES. 50, 275 (1979)

Thompson, J. M. T., Instabilities and Catastrophes in Science and Engineering, John Wiley, New York (1982)

Thompson, J. M. T., Bishop, S. R., and Leung, L. M., Phys. Lett. 121A, 116 (1987)

Thompson, J. M. T., and Ghaffari, R., Phys. Lett. 91A, 5 (1982)

Thompson, J. M. T., Bokaian, A. R., and Ghaffari, R., IMA J. Appl. Math. 31, 207 (1983)

Thompson, J. M. T., and Ghaffari, R., Phys. Rev. A27, 1741 (1983)

Thompson, J. M. T., and Stewart, H. B., Nonlinear Dynamics and Chaos, John Wiley, New York (1986)

Tomita, K., and Kai, T., J. Stat. Phys. 21, 65 (1979)

Tomita, K., and Daido, H., Phys. Lett. 79A, 133 (1980)

Tomita, K., Phys. Reports 86, 113 (1982)

Tresser, C., Coullet, P., and Arneodo, A., J. de Phys. Lett. 41, L243 (1980)

Tylikowski, A., and Skalmierski, B., Stochastic Processes in Dynamics, PWN, Warsaw (1982)

Ueda, Y., J. Stat. Phys. 20, 181 (1979)

Ueda, Y., Ann. N. Y. Acad. Sci. 357, 422 (1980)

Ueda, Y., and Akamatsu, N., IEEE Trans. Circuits and Systems CAS28, 217 (1981)

Ueda, Y., and Aizawa, Y., Prog. Theor. Phys. 68, 1543 (1982)

Ueda, Y., and Ohta, H., Proc. ISCAS 85, 179 (1985)

Ueda, Y., Survey of Strange Attractors and Chaotically Transitional Phenomena in the System Governed by Duffing's Equation, in Complex and Distributed Systems: Analysis, Simulation and Control ed. Tzafestas, S. G., and Borne, P., Elsevier, Amsterdam (1986)

Vallee, R., Delisle, C., and Chrostowski, J., Phys. Rev. A30, 336 (1984)

Wolf, A., and Swift, J., Phys. Lett. 83A, 184 (1981)

Sano, S., Sawada, Y., Phys. Lett. 97A, 73 (1983)
Sato, S., Sano, M., and Sawada, Y., Phys. Rev. A28, 1654 (1983)
Schulmann, N. J., Phys. Rev. A28, 477 (1983)
Shinczuka, M., J. Acoustical Soc. America 49, 357 (1977)
Shaiman, B., Wayne, C. E., and Martin, P. C., Phys. Rev. Lett. 46, 935 (1981)
Shaw, S. W., and Holmes, P., Phys. Rev. Lett. 51, 623 (1983)
Shaw, R. S., Z. Naturforsch. 36a, 80 (1981)
Shimada, I., and Nagashima, T., Prog. Theor.Phys. 61, 1605 (1979)
Shimada, I., and Nagashima, T., Prog. Theor. Phys. 59, 1033 (1978)
Steeb, W.-H., Erig, W., and Kunick, A., Phys. Lett. 93A, 267 (1983)
Steeb, W.-H., Louw, J. A., Leach, P. G. L., and Mahomed, F., Phys. Rev. A33, 2131 (1986)
Steeb, W.-H., Louw, J. A., and Kapitaniak, T., J. Phys. Soc. Jpn. 55, 3279 (1986)
Steeb, W.-H., and Kunick, A., J. Phys. Soc. Jpn. 54, 1220 (1985)
Steeb, W.-H., and Kunick, A., Int. J. Non-linear Mech. 23 (1987)
Steeb, W.-H., and Louw, J. A., Chaos and Quantum Chaos, World Scientific, Singapore (1986)
Stratonovitch, R. L., Topics in the Theory of Random Noise, Gordon and Breach, New York (1963)
Szemplińska-Stupnicka, W., IFTR Reports 27, (1986)
Szemplińska-Stupnicka, W., and Bajkowski, J., IFTR Reports 4, (1986)
Szopa, J., Zeszyty Naukowe PS, Math.-Phys. 45, (1985)
Szopa, J., J. Sound Vibr. 97 , 645 (1984)
Szopa, J., and Bestle, D., J. Sound Vibr. 107, 177 (1986)
Tagata, G., J. Sound Vibr. 58, 95 (1978)
Thomas, S., and Grossmann, S., J. Stat. Phys. 26, 485 (1981)

Wolf, A., Quantifying Chaos with Lyapunov Exponents, in Chaos, ed. Holden, A. V., MUP, Manchester (1986)

Wolf, A., Swift, J. B., Swinney, H. L., and Vastano, J. A., Physica 16D, 285 (1985)

Wright, J., Phys. Rev. A29, 2924 (1984)

Wróbel, J., Prace Naukowe PW, Mechanika 92, (1985)

Yorke, J. A., and Yorke, E. D., J. Stat. Phys. 21, 263 (1979)

Zardecki, A., Phys. Lett. 90A, 274 (1982)

Zippelius, A., and Lucke, M., J. Stat. Phys. 24, 345 (1981)

Zgone, K., and Grabec, I., Chaotic Movement of a Simple Mechanical Machine, Preprint , (1987)